EU CLIMATE POLICY

EU Climate Policy
Industry, Policy Interaction and External Environment

ELIN LERUM BOASSON
CICERO Center for Environmental and Climate Research, Norway

JØRGEN WETTESTAD
Fridtjof Nansen Institute, Norway

Routledge
Taylor & Francis Group

LONDON AND NEW YORK

First published 2013 by Ashgate Publishing

Published 2016 by Routledge
2 Park Square, Milton Park, Abingdon, Oxon OX14 4RN
711 Third Avenue, New York, NY 10017, USA

First issued in paperback 2017

Routledge is an imprint of the Taylor & Francis Group, an informa business

British Library Cataloguing in Publication Data
Boasson, Elin Lerum, 1978–
 EU climate policy : industry, policy interaction and external environment.
 1. Climatic changes—Government policy—European Union countries.
 2. Environmental policy—European Union countries. 3. Energy policy—
 European Union countries. 4. Business and politics—European Union countries.
 I. Title II. Wettestad, Jørgen, 1955–
 363.7'056'094–dc23

Library of Congress Cataloging-in-Publication Data
Boasson, Elin Lerum, 1978–
 EU climate policy : industry, policy interaction, and external environment / by Elin Lerum Boasson and Jørgen Wettestad.
  p. cm.
 Includes bibliographical references and index.
 ISBN 978-1-4094-0355-5 (hardback : alk. paper)
 1. Energy policy—European Union countries. 2. Climatic changes—Government policy—
 European Union countries. I. Wettestad, Jørgen, 1955– II. Title.
 HD9502.E82B63 2013
 363.738'74561094—dc23

                                                                    2012022206

ISBN 13: 978-1-138-05115-7 (pbk)
ISBN 13: 978-1-4094-0355-5 (hbk)

Contents

List of Tables

List of Abbreviations

ALTENER	Programme for the promotion of renewable sources of energy in the European Community
BP	British Petroleum
CAN	Climate Action Network
CCS	Carbon Capture and Storage
CDM	Clean Development Mechanism (UNFCCC)
CECHODHAS	The federation of public, cooperative and social housing
CEEC	Central and Eastern European Countries
CEN	European Committee for Standardization
CO2	Carbon Dioxide
COP	Conference of the Parties (UNFCCC)
DG	Directorate (within the European Commission)
ECCP	European Climate Change Programme
EC	European Community
EPBD	Energy Performance of Buildings Directive
EREC	European Renewable Energy Council
E3G	British environmental organization
ETP	European Technology Platform
ETS	Emissions Trading System
EU	European Union
EUFORES	European Forum for Renewable Energy Sources
EURELECTRIC	The Union of the Electricity Industry
EURIMA	Insulation trade association
EuroACE	European Alliance of Companies for Energy Efficiency in Buildings
EUROFER	European steel association
FIEC	European Construction Industry Federation
GETS	Greenhouse Gas and Energy Trading Simulations
GDP	Gross Domestic Product
GHG	Greenhouse Gases
ICSU	International Council for Science
IEA	International Energy Agency
IPCC	Intergovernmental Panel on Climate Change
IPPC	Integrated Pollution Prevention and Control (EU Directive)
JI	Joint Implementation (UNFCCC)
LI	Liberal Intergovernmentalism
MEP	Member of European Parliament
MLG	Multi-level Governance

NAP	National Allocation Plan (the ETS)
NER	New Entrants Reserve (the ETS)
NGO	Non-Governmental Organization
NI	New Institutionalism
OECD	Organization for Economic Co-operation and Development
OSPAR	Oslo and Paris Conventions
R&D	Research and Development
RES	Renewable energy
SAVE	Specific Actions for Vigorous Energy Efficiency (EU Programme)
SEA	Single European Act
SET	Strategic Energy Technology Plan
TwH	Terawatt Hours
UK	United Kingdom
UNCCC	United Nations Framework Convention on Climate Change
UNEP	United Nations Environment Programme
WMO	World Meteorological Organization
WWF	World Wildlife Fund
ZEP	Zero Emission Fossil Fuel Power Plants (ETP)

Preface

Work on this book was initiated by Jørgen in late 2008, in response to the finalization of the EU Climate and Energy package at the time. There was a need for something that could cover and compare several key policies in the 'new climate policy drive' and probe into the main driving forces behind the development of such policy. But it was when Elin joined the project in the spring of 2009 that things really got going. Elin contributed to expanding the ambitions of the book significantly in terms of theory and method, not least the importance of identifying core mechanisms. She also brought in the little-studied case of energy performance of buildings.

The book is truly a result of collaborative efforts: we have worked together on all chapters. This cooperation worked well and eventually turned into a more stable joint venture. As the incredible jewel in the crown of that process, our daughter Liv was born on 22 September 2011. Of course her arrival made the process of book finalization somewhat more complex – but so much more joyful!

We wish to extend our warm thanks to many individuals and groups of people, presented here in roughly alphabetical order:

> Michelle Betsill, for the wonderfully encouraging response at the ISA Convention in 2010;
> Anne Raaum Christensen, for helpful comments on Chapters 4 and 5;
> Tom Christensen, for comments on an early comparative draft, which helped us to improve the structure of the assessment;
> Per Ove Eikeland, for helpful comments on Chapter 5;
> Neil Fligstein, and other participants at a spring 2010 workshop at the Berkeley Center for Culture Organization and Politics, for fruitful comments on an early comparative draft – especially useful for our work on the new institutional dimensions and long-term historical assessment of the cases;
> Paal Frisvold, for helpful comments and inputs to the work on Chapter 6;
> Susan Høivik, for excellent language polishing, general encouragement and for making the finalization process more enjoyable;
> Geir Hønneland, for believing in this project from the very start and providing encouragement in the crucial initial stages;
> Tor Håkon Inderberg, for helpful comments to Chapters 4 and 5, and general encouragement along the way;
> Liv Arntzen Løchen, for very useful research assistance, not least in connection with Chapter 3;

Sebastian Oberthür, for reviewing a previous version of the manuscript and providing a thorough, thoughtful and extremely useful review report;

Maryanne Rygg, for careful and meticulous editing assistance;

Peter Johan Schei and Arild Moe, for allocating additional FNI funding for work on this book project;

Rob Sorsby and the rest of the Ashgate team, for the book contract and for their generosity in allowing us extra time to finish the work;

Olav Schram Stokke, for reviewing a previous version of the manuscript and providing generous, insightful comments;

Jarle Trondal, for reviewing a previous version of the manuscript and helping us to relate our work better to EU integration theory;

Arild Underdal, for very useful comments on our 2011 ISA paper, helping us to tighten our analytical framework;

Inga Ydersbond, for helpful comments on Chapter 5; and

John Zysman for helpful comments on an early comparative draft.

Our thanks also to the anonymous reviewer provided by Ashgate, who encouraged us to write a book that would not be 'so quickly dated'. We hope that we have been able to do so!

Oslo, March 2012
Elin Lerum Boasson and Jørgen Wettestad

Chapter 1

Introduction

Introduction

EU climate and energy policy was once a matter of minor political importance on the EU agenda, with issues like the internal market, enlargement and monetary policies ranking far higher. After a decade of severe political conflicts and primarily symbolic policy outcomes in the 1990s, climate policy soared from 'just another' part of EU environmental policy to become a high-profile policy area in its own right. Especially from 2005 onwards, the pace of developments has been rapid indeed. By the end of that decade, a range of new and ambitious targets had been adopted, complemented by a broad palette of tangible and binding policies. Climate policy has emerged as a vital area of EU governance.

Ups and downs characterize the development of environmental policy, as noted by Anthony Downs already back in 1972 (Downs 1972). The 'new drive' in EU climate policy has been followed by financial crisis; as a result, issues of climate change and the related challenges are deemed less urgent, and have slipped further down on the political agenda.

This book discusses the central dynamics in the emergence and development of EU climate policy, and the driving forces behind it. Our overarching research question is: *how can we best explain the development of EU climate policy development?* We aim to shed new light on this, by taking into account both long-term historical developments and the impressive drive of recent years. What emerges is a fascinating story, not least of skilled entrepreneurs who have managed to create and exploit political windows of opportunity (see also Kingdon [1984] 2011). There are examples of more instant '*carpe diem*' entrepreneurship as well as more persistent and long-term 'tortoise' entrepreneurship. This is also a story of institutional feedback processes. To capture this, we combine sociological New Institutionalism and political science theories in a novel way. Drawing on information from more than 60 interviewees, including both central actors and observers, we present stories never told before.

The book provides structured, comparative case studies of the emergence and development of four central climate sub-policies: the emissions trading system (ETS), renewables, carbon capture and storage (CCS), and energy policy for buildings. There are intriguing differences in the characters of these four. Whereas the ETS creates a basically harmonized European market for greenhouse gas permits, only indirectly inducing technology choices and development, the EU has involved itself rather directly in renewable energy and in CCS, aiming to transform industry in these areas. And the EU energy policy for buildings provides a framework for regulation of national building construction, giving member states significant leeway to determine specific technology standards. To capture the variation, we provide a new typology of differing EU climate policy outcomes.

European integration theory has prospered in recent decades, developing a rich body of literature that draws on a wide range of approaches. This offers a good foundation for understanding EU climate policy development. We develop an explanatory framework that relies primarily on three approaches to European integration: Liberal Intergovernmentalism (LI), Multi-level Governance (MLG) and New Institutionalism (NI). These perspectives can yield valuable insights, but none of them have been specifically developed to explain climate policy. Rather, they guide our conceptualization of the key mechanisms behind the development of climate policy. Here we seek to contribute to theory development in this field through the explicit formulation and empirical application of these mechanisms.

EU policymaking is a highly complex and multi-level venture, so it would be foolhardy to try to cover 'everything'. Taking into account previous studies of EU climate policy (see the overview in Chapter 3), we have selected three main topics of theoretical interest and practical importance for special attention throughout the book:

- first, the role of industry, as a central stakeholder, target group, lobbyist and implementer of adopted policies;
- second, the role of policy interaction, since the 'package' concept has figured centrally in the new drive in EU climate policy and the European Commission (hereafter: Commission) has argued that EU climate policy now consists of well integrated sub-policies;
- third, the role of the EU-external environment, as many EU actors (the Commission not least) have long seen EU policy development as closely related to international climate negotiations and international energy prices.

On this backdrop we have made a deliberate selection of cases. We focus on cases characterized by varying likelihood that industry, policy interaction and the EU-external environment have played significant roles (see George and Bennett 2005, Gerring 2008). We base our assessments on extensive reviews of the documentation, as well as information and insights gained from over 60 interviewees. The combination of extensive empirical material, comparative case technique and active use of theory provides a good foundation for understanding how EU climate policy has developed thus far, as well as enabling some educated guesses about future policy developments. By no means is this book meant as the final say on the mechanisms that shape EU climate policy, and we invite fellow political scientists as well as EU policymakers to respond to our conclusions.

Capturing EU Climate Policy Variation: Steering Method and Competence Distribution

Climate policy has developed into a distinct and important policy area in the EU, but the character of its various sub-areas varies significantly. In order to understand how EU climate policy develops, we first need to capture this variation analytically. Scholarly discussions on policy conceptualizations have been central to the emergence of common and fruitful research programmes relating to welfare states, democracies, revolutions, and economic organization (see Hall and Soskice 2001, Mahoney and Rueschemeyer 2003). Shared conceptualizations can hinder misunderstandings and misinterpretations, and enable researchers from different traditions to trace patterns of commonalities and variation across a large number of cases – in turn, fostering fruitful scientific discussion.

EU policy discussions often centre on to two dimensions: the Steering Method and the Competence Distribution (see Wallace and Wallace 2006, Young 2006). Most actors involved in climate policy debates agree that low-carbon technologies and industry practices will have to become more profitable, and that new and better technologies and technical practices are needed (Metz et al. 2007). However, they do not agree on how to achieve this, or the steering methods. Some favour market measures that put a price on carbon emissions or in other ways may indirectly induce technology development, whereas others favour technology development measures focused more directly on ensuring that industry will develop and make use of specific technologies.

Market measures either create new markets or alter existing ones. They may put a price on polluting activities (as with emissions trading) or reduce the price/generate an extra income on low-carbon activities (like green certificate markets) or bring new information into price-forming processes (like quality labels and certificates). The rationale is that the regulated industries will develop low-carbon practices once this is economically viable (Sims et al. 2007: 306). The prime task of governments and/or the EU is then to design markets that can make it economically beneficial to produce and use low-carbon products. Market measures do not in principle favour any specific industries or technologies, but allow the market forces to choose winner industries and technologies. The precise economic incentives will not be set by the EU or the governments, but are to be shaped by the market.

In contrast, technology development steering may involve technological standards (such as building codes), technology-specific governmental support (such as feed-in schemes) or governmental investments in certain technologies. It is assumed that low-carbon technologies and practices will diffuse once the technologies are mature and technology competencies have become widely disseminated. The prime task of governments and/or the EU is then to develop technology standards, and to finance R&D, demonstration project and training activities. Governments are advised to apply a wide range of measures, adjusted to the specific needs of different industries and the various low-carbon products that are under development (Sims et al. 2007: 306).

The second dimension, the Competence Distribution, goes to the core of European integration theory, and has been described by Haas as 'the shift of loyalties, expectations and political activities toward a new centre, whose institutions possess or demand jurisdiction over the pre-existing national states' (Haas [1958] 2004: 16). This broad definition encompasses both cultural and structural centralization, and the increased number of EU policies and EU member states. We will focus more specifically on the distribution of authority within our selected climate-policy areas. A key point here is whether it is the member states or the EU organizations that are given the basic competence to govern the policy issue in question (Olsen 2007: 96).

Scholars have employed a range of terms to characterize this kind of centralization of power within the EU. For instance, Schimmelfenning and Rittberger (2006: 74–75) use the term 'vertical integration' in describing the distribution of competencies between EU organizations and member states. The more power that EU organizations are given to govern 'joint decision-making, implementation and enforcement', the stronger will be the degree of centralization (Moravcsik 1993: 479). This relates in particular to issues such as whether the EU or national governments have competence to change basic features of the policy, for instance by developing detailed regulations; and issues of monitoring of progress and implementation in the policy area. EU organizations can be seen as stronger if they do this directly, than if they rely on information supplied by member states, and have competence to ensure enforcement by use of coercive measures, like formal infringement cases and fines.

Often, policies will not be clear-cut, and may contain differing elements. Still, we believe that it will normally be possible to identify which steering method and competence distribution dominates a certain policy outcome. Combinations of the two dimensions create the four categories presented in the two-by-two table, Table 1.1.

Table 1.1   Four types of EU climate policies

Competence distribution	Steering method	
	Technological development	*Market approaches*
National/'decentralized'	1) Local loading	2) Piecemeal market
EU governance	3) EU engineering	4) Single European market

'Local Loading' EU policies can be seen as giving member states the upper hand, and encouraging the use of national technology development instruments, ranging from state aid schemes directed at promoting specific technologies, to emission limits and other governmental regulations. Also what we refer to as 'Piecemeal

Market' policies give member states most of the competence to develop policies, while encouraging the use of national market instruments. Hence, here we will find market change at the national level, but no European harmonization of markets. What we call 'EU Engineering' policies give EU competence to engage directly in promoting specific technologies, whether through technological requirements or funding of specific technologies. Finally, 'Single European Market' policies give the EU the upper hand in the creation, surveillance and control of EU-wide market measures. This latter outcome is often characterized as 'harmonized EU policy' and has been in focus among scholars of European integration. Neither climate policy stakeholders nor analysts agree on whether this policy outcome is the best way towards a low-carbon society, however.

We will pay special attention to how Single European Market outcomes are created, but we do not see this outcome as normatively superior – such judgements are left to the reader. In the following, we present the theory basis for each of our three theory topics.

Exploring the Role of Industry: Simple Braking Block – or More Multi-faceted Player?

Media accounts and empirical assessments of Brussels climate-policy processes tend to attribute considerable power and influence to industry lobbyists. For instance the scholarly accounts of the failed efforts of establishing a carbon tax in the 1990s highlight the braking block role played by industry (Skjærseth 1994, Ikwue and Skea 1994). Other pieces of evidence point in different directions, for instance highlighting the role of big oil companies as emissions trading frontrunners and inspiration sources for the subsequent EU policy development (see Christiansen and Wettestad 2003, Victor and House 2006). What role does industry play for the emergence and development of EU climate policy – is it a simple braking block, easily analytically captured? Or is it a far more multi-faceted player, with differing roles in differing phases and issues? Can mechanisms derived from the theory perspectives help us to pin down the actual role played by industry?

The three perspectives provide different propositions in this respect. Liberal Intergovernmentalists (LIs) posit that national governments will act as instruments for economically dominant national industries (Moravcsik 1993, 1998). New Institutionalists (NIs) highlight the mutual interdependence between industry and governments, and the common culture and institutional features they have developed over time (Olsen 2007, Fligstein 2008). Multi-level Governance (MLG) scholars argue that shrewd Brussels entrepreneurs able to create multi-level networks (spanning national and EU levels of policymaking) will be the most influential (Coen 1998, Mazey and Richardson 2006, Marks, Hooghe and Blank 1996). We will explore whether the three perspectives in conjunction are able to grasp the most important mechanisms through which industry affects EU climate policy, and will then develop some proposals for analytical improvements.

The historical process-tracing of the empirical studies enables us to follow and discuss the long-term consequences of industry influence, not only the more temporary effects. The four climate policy cases target industries with varying business practices, differing Brussels presence, and varying degrees of transnationalization. Further, included in the sample are two cases where earlier studies have indicated that industry played a major role: the ETS and Renewables; as well as two cases where industry seems to have been less active: CCS and Buildings (Skjærseth and Wettestad 2008, Boasson 2011). On this backdrop we will discuss whether traditional European integration approaches can provide appropriate conceptualizations of the impact of industry impact. Further we will probe into whether there is any clear relationship between the character of the industries involved in policymaking and the four types of climate policy outcomes.

Exploring the Role of Policy Interaction: Actual Mingling and Cross-fertilization – or 'Isolated Icebergs'?

Different EU policies may affect each other. The first wave of European integration theory, neofunctionalism, attributed considerable explanatory power to policy interaction, or spillover, which was the term used by Ernst B. Haas ([1958] 2004). Haas was particularly interested in interaction over time, and predicted that issue-areas in which the EU gained considerable competence would increasingly affect other policy areas. International environmental politics scholars have paid increasing attention to interaction, and conclude that the EU's various environmental policies have significant impact on each other (cf. Oberthür and Gehring 2006). In contrast, recent studies of EU administration have found fairly little communication across sectors or policy-areas (Egeberg 2006a, Trondal 2010).

What role, then, has policy interaction played in the emergence and development of EU climate policy? Has 'mingling' and cross-fertilization between policies led to radically different policy outcomes? Or will closer examination show that policies have developed primarily due to issue-internal factors, as 'isolated icebergs'? Can mechanisms derived from our theory perspectives help us in pinning down the actual role of policy interaction?

In LI, particular attention is paid to how different policies may be interlinked through bargained deals (Moravcsik 1998). This happens because national officials try to link issues in a way that enhance their clout. That perspective will also make it natural to expect that policies may have functional and economic consequences, such as market changes, for other policy areas. By contrast, the NI approach focuses on how the policy recipe, norms, practices or worldviews that underpin one policy may contribute to shape other policies (see Fligstein and Stone Sweet 2002). And thirdly, although the MLG approach does not discuss interaction very explicitly, it is in line with the thinking of this perspective to expect EU officials to initiate linkages between policy areas.

We will conduct long-term process tracing in order to identify when interaction occurs and how it affects policymaking over time. As noted, we have two cases where previous research indicates that interaction played a major role for the policy outcome (CCS and energy policy for buildings), and two where issue-internal dynamics might have been more important (the ETS, and renewables) (see Claes and Frisvold 2009, Wettestad, Eikeland and Nilsson 2012). Can existing theories of European integration capture the major policy interaction effects? And is there a relation between policy interaction and the type of climate policy outcome?

Exploring the Role of the External Environment: Binding Hands – or Handy Opportunities?

Climate change has become an important issue in international politics. This is indeed a policy area in which EU policymaking is clearly linked to EU-external factors, and in particular the rules, practices and negotiations of the global climate regime (see Oberthür 2006, Skjærseth and Wettestad 2008, Wettestad 2009c). Also relevant are the global and regional energy situation and international market developments for various energy sources and technologies (Jordan and Rayner 2010). Thus we need to explore how factors external to the EU have played into policymaking (Lenschow 2006: 425).

What role has the external environment played for the emergence and development of EU climate policy? Have such concerns bound the hands of EU policymakers, giving them scant leeway in the policy development process? Or have developments in the external environment primarily created convenient opportunities for creative policy entrepreneurs who have used the international backdrop to promote their own ideas and interests? How can mechanisms derived from our perspectives help us identify the actual role of the EU-external environment?

Although most EU policies have an external policy dimension, such EU-external factors have not been much discussed in European integration studies (Jørgensen 2006: 511). Instead, the European community is often portrayed as a sort of 'closed system' affected solely by EU-internal developments. LI does not explicitly discuss the potential policy effect of EU-external factors, but it follows the logic of the perspective to expect that changes in international market conditions may change the positions of national governments. Also most NI approaches to EU integration have concentrated on internal factors, but the broader NI literature draws attention to how local institutional orders and policies may be undermined, strengthened or changed due to international developments (Meyer et al. 1997, Drori, Meyer and Hwang 2006). Hence, this perspective leads us to expect that policy recipes adopted by other countries or environmental regimes will diffuse and inspire EU policymakers. Thirdly, MLG scholars pay little attention to the external environment, but the very idea of multi-level entrepreneurship points to the possibility that EU actors may use global arenas strategically in order to increase their influence on internal processes within the EU.

We will explore whether any of these perspectives can shed light on how external features have affected EU climate policy outcomes. The external environment may influence all types of climate policies, but prior research indicates that the ETS and CCS policies have been more affected than have EU policies for renewable energy and buildings (Claes and Frisvold 2009, Boasson 2011, Wettestad, Eikeland and Nilsson 2012). Such historical case studies make it possible for us to discuss the long-term consequences of EU-external influence, and not only more temporal effects. We will also examine whether there is a relationship between EU-external influence and the character of climate policy outcomes.

Outline of the Book

The next chapter presents the theoretical and methodological framework. Chapter 3 provides a summary overview of the development of EU climate policy, emphasizing key developments at the EU level, and the relationship to external developments. Chapters 4 to 7 present the cases in focus: Emissions Trading (Chapter 4); Renewables (Chapter 5); Carbon Capture and Storage (Chapter 6); and Energy Policy for Buildings (Chapter 7). Chapter 8 then offers a comparative assessment of the four cases and presents final conclusions and theory implications.

Chapter 2

Theory and Method

Introduction

This chapter provides a foundation, in theory and method, for understanding EU policy development. We focus on three topics of high theoretical interest and practical importance for EU climate policy: to what extent and how industry, EU policy interaction and the external environment contribute to shape EU policy outcomes. As noted, the literature on European integration has paid scant attention to these issues. Hence, analytical clarification of the three themes is important for the work to be carried out in this book and beyond.

We will apply three dominant European integration theory approaches: Liberal Intergovernmentalism (LI), New Institutionalism (NI) and Multi-level Governance (MLG). The three draw on different traditions within the social sciences. Liberal Intergovernmentalism builds on neo-realism and rational choice. The term 'New Institutionalism' has come to cover a range of different approaches to European integration: here we will draw on both sociological and historical NI contributions. Multi-level Governance is a looser theoretical tradition, inspired by pluralist and network approaches to national policymaking. In addition to the three specific literatures proper, we also draw on the larger theory communities of which they are part. This rich theory backdrop will be employed to specify policy-shaping *mechanisms*. Such 'mechanisms' are descriptions of the actual paths or features that connect the three causes (industry, policy interaction and the external environment) to the effects (the policy outcomes) (Hedström 2008).

Our specific focus is on two central dimensions of EU climate policy outcomes: the distribution of competences, and the steering method. All perspectives discuss, more or less explicitly, why EU member states delegate power to EU, thereby increasing the centralization of control. The steering method issue has received less explicit attention, but the theories can nonetheless help us to identify different causal drivers.

The Three Basic Approaches: Liberal Intergovernmentalism, New Institutionalism and Multi-level Governance

Liberal Intergovernmentalism: Rational National Governments Defending Industry Interests

Andrew Moravcsik has developed Liberal Intergovernmentalism (LI) by drawing on neo-realism, rational choice approaches (also called rational institutionalism), traditional intergovernmentalism and bargaining theory. This is a 'grand theory' with coherent and neat specifications of a set of policy shaping mechanisms (Pollack 2001). It is assumed that only states will have direct influence on EU policy outcomes, and their preferences will be shaped by national conditions that reflect the economic interests of economically dominant national constituents (Moravcsik and Schimmelfenning 2009: 68). Dominant member states will have the greatest influence on EU policy outcomes (Moravcsik 1993, 1998, Moravcsik and Schimmelfenning 2009: 68).

> EU integration can best be understood as a series of rational choices made by national leaders. These choices responded to constraints and opportunities stemming from the economic interests of powerful domestic constituents, the relative power of states stemming from asymmetrical interdependence, and the role of institutions in bolstering the credibility of interstate commitments.
> (Moravcsik 1998: 18)

EU decision-making is pictured as a three-stage process. First, national positions will reflect the bargaining power of economically dominant national actors, which in practice means the most dominant industries (since other societal actors seldom have significant economic power) (Moravcsik 1993: 483, 1998: 18). Note that Moravcsik has argued that it is a common misunderstanding that LI's basic claim is that 'producer interests prevail' or 'economics dominate policy' (Moravcsik and Schimmelfenning 2009: 70). At the same time he emphasizes that 'the preferences of national governments regarding European integration have mainly reflected concrete economic interests rather than other general concerns' and repeats his earlier arguments on the importance of nationally powerful economic actors (Moravcsik and Schimmelfenning 2009: 70). Moravcsik does not discuss under which conditions member states are most likely to support market measures and when they may be inclined to support technology development steering, but he specifically argues that governments will seek multilateral trade liberalization when it is no longer possible to realize producer interests unilaterally (Moravcsik 1998: 38).

Second, the member states will defend the interests of economic dominant national industries in EU-level bargains. Because information and ideas are seen as plentiful and assumed to be relatively symmetrically distributed among states, supranational actors (whether the Commission or other EU actors) have little influence (Moravcsik and Schimmelfenning 2009: 71). Here, the member states least in need of a specific outcome, relative to the status quo, are the best able to threaten others into making concessions.

Third, member states will delegate authority to EU bodies if that is needed in order to realize and implement the political agreement. This implies that member states will generally be reluctant towards centralized EU policy; however, if this is needed to ensure a well-functioning EU policy, they may make concessions. For instance, they will agree to centralize authority if this is needed in order to hinder 'the build-up of national pressure for non-compliance if costs for powerful domestic actors are high' (Moravcsik 1998: 9, 73).

The LI approach posits that member-state preferences are shaped through a bottom-up process: governments act as instruments for their economically dominant national industries, and member states have the chief say in EU decision-making. On those rare occasions where producer interests are weak, diffuse or indeterminate, other concerns may flavour member-state positions (Moravcsik 1998: 7). The environment is mentioned as an area in which politicians tend to have 'independent preferences for regulatory goals' (Moravcsik 1998: 37). Hence, Moravcsik admits that other factors may be important, but provides no theoretical specification of such alternative preference formation. EU climate policy is more directly related to the economic interests of industries than are many other environmental policies. This makes climate policy particularly well-suited for testing the LI approach.

New Institutionalism: Organizational Fields with Specific Historical Feedback Effects

Several social science approaches go under the name 'New Institutionalism' (NI). Here, we present an approach to European integration that draws mainly on sociological and historical NI, but also on certain versions of neo-functionalism. While other approaches to European integration focus on the political process proper, this perspective takes into account the wider organization of the economy and patterns of social actions. Prominent scholar in this tradition Neil Fligstein (2008: 9) notes: 'the EU is like an iceberg: what goes on in Brussels is like the 10 per cent above the waterline. But the really interesting story is the 90 per cent that is harder to see'.

Such a perspective gives prime attention to culture and institutional phenomena, such as norms and conventions, that guide interest formation and policy development. These features vary among different European sub-communities or spheres with specific societal orders (see DiMaggio and Powell 1991, Scott 2008). These sub-communities may be termed *organizational fields*, defined by Fligstein (2008: 8) as:

> an arena of social interaction where organized individuals or groups such as interest groups, states, firms, and non-governmental organizations routinely interact under a set of shared understandings about the nature of the goal of the field, the rules governing social interaction, who has the power and why and how actors make sense of anothers' actions.

Organizational fields are characterized by a symbiotic relationship between cultural-institutional phenomena, EU organizations, governments and industries (Fligstein and Stone Sweet 2002: 1207). Industry, governments and EU organizations involved in the same field tend to be mutually interdependent. The

EU is seen as a multi-tiered, multi-structured and multi-centred polity, involved in an array of organizational fields (Olsen 2007). The institutional character of the organizational field will flavour the EU policy outcomes. Sociologist Richard Scott (2008: 48) defines institutions as 'comprised of regulative, normative and cultural-cognitive elements that, together with associated activities and resources, provide stability and meaning to social life'. The regulative structures consist of formal rules that involve coercive sanctions, whereas the normative and cognitive elements are backed up by social, informal reactions. We will particularly focus on the normative elements (values, norms and moral principles) and cognitive elements – the latter which Scott (2008: 57) describes as 'the shared conceptions that constitutes the nature of social reality and the frames through which meaning is created'.

The concept of *institutional logic* builds on the term 'logic of appropriateness' developed by March and Olsen (1989, 1998). They suggest that actors will act on the basis of how they understand their role, how they interpret the situations they face, and which rules they regard as appropriate to follow in such circumstances. Friedland and Alford (1991: 240) have taken this approach one step further by suggesting that there is a systematic relationship between identities and action. Thornton (2004: 42, see also Thornton and Ocasio 2008) has conceptualized institutional logics as specifying *a priori* ideal-type models of cultural practice and symbol systems that provide fairly coherent templates for how to act in different situations. Such logics will determine which issues and problems to attend to and what answers and solutions are deemed most appropriate (Thornton 2004: 13–14).

Climate policy debates are often characterized by conflict between the market logic of business economists on the one hand and the technology development logic of engineers on the other. Key elements of both logics are presented in table 2.1. The market logic posits that commercial organizations possess perfect information and are capable of acting strategically on this information (Fligstein 2001: 13). It is further assumed that firms will strive to maximize their profits in a medium- to long-term perspective. According to this logic, governments should work to ensure that low-carbon investments are the most profitable ones, and support schemes should encourage market actors to compete in developing the most profitable projects (Sims et al. 2007: 306). Thereby the 'best' projects will be developed, and actors able to develop the most profitable projects will be rewarded with the greatest profits. Basically, both emissions trading systems as well as green certificate schemes for renewable energy have been designed in line with this logic (see European Commission 2005a, 2008j).

The technological development logic is based on technological rather than economic criteria. It is assumed that industrial change hinges on technological innovation and its subsequent refinement. Commercial organizations will aim to enhance technological development, and governments should ensure good and stable conditions enabling them to do so. It is the technical quality of the alternatives to conventional production that will determine the levels of support: different technologies will receive different levels of support. Moreover, support schemes will be designed to ensure long-term stability, so that commercial actors may expend the time and resources needed to refine those technologies in which they have greatest expertise (Sims et al. 2007: 306). Feed-in tariff schemes that guarantee renewable-energy producers access to the grid and a fixed level of operational support, and varying levels of support for different technologies, fit this approach well (European Commission 2005a, 2008j), as do technology standards like emission limits or energy performance requirements.

Table 2.1   Main institutional logics underpinning climate-policy development

Logic components	Type of logic	
	Market logic	*Technical development*
Objective	Maximizing corporate profits	Enhancing technical development
Role of commercial organizations	Engaging in strategic competition aimed at maximizing corporate profits	Inventing, developing and refining promising technologies
Appropriate solutions	Market-based support schemes that favour the most profitable low-carbon solutions	Fostering a wide spectrum of technologies by introducing various technology-specific measures
Governmental measures	Market measures (e.g. emissions trading or green certificate schemes)	Feed-in schemes. Technology standards

Organizational fields dominated by technology development logics are prone to favour technology development policy outcomes, whereas fields characterized by logics tend to produce market measures. The existence of multiple institutional logics enables actors to challenge existing policies as well as propose new policy options (Friedland and Alford 1991: 254, Lounsbury 2007: 302, Reay and Hinings 2009). For instance, an actor who is dissatisfied with a policy and its impact will be strengthened if that actor can argue that the policy runs counter to a dominant logic, and can then proceed to propose a new policy measure that fits the dominant logic better. Different organizational fields may be embedded in different institutional logics, and may hence prescribe different policy recipes.

In addition, the structure of the organizational fields will vary, in the sense that distribution of authority between national actors and European actors may vary (Fligstein 2008). The structure of the organizational fields will contribute to shape the competence distribution of EU policies.

Because different organizational fields are involved in different EU policy processes, we will see different policy outcomes (see Fligstein 2008). Moreover, each field will create differing institutional feedback mechanisms that over time will shape the steering method and competence distribution of EU policies (Pierson 1996, Fligstein 2008, see also Niemann 2006). Each period in the evolution of a policy will contribute to the creation and specification of the institutional logic(s) and authority distribution, structuring 'what takes place thereafter' (Fligstein and Stone Sweet 2002: 1230).

Everyday policymaking and the resolution of technical questions will contribute to specify logics and eventually create taken-for-granted understandings of appropriate steering methods and competence distribution (Pierson 1996: 130, Lawrence and Suddaby 2006, Scott 2008). Shifts in logics or competence distribution may alter how actors 'pursue their interests, how national judiciaries operate, and how policy is made' (Fligstein and Stone Sweet 2002: 1219). As social actors make commitments based on existing institutions and policies within their fields, any exit from existing arrangements becomes unlikely (Pierson 1996: 146). Moreover, the room for agreement among EU member states with diverse models of governance is expected to be significantly smaller in policy areas traditionally governed by the member states, than in the new domains where historical policy ideas are not as deeply entrenched (Fioretos 2011: 385).

Basically then, according to the NI perspective, member-state positions will be shaped by long-term institutional feedback mechanisms of the organizational fields operative within a certain policy area. The institutional logic of the organizational field(s) will contribute to shape the steering method of the policy outcome, whereas the pre-existing distribution of competence within the field will shape whether new policy outcomes end up as decentralized or centralized. The structure of the organizational fields will help to shape whether top-down or bottom-up preference formation dominates. Europeanized fields will tend to be characterized by top-down preference formation: the positions of national governments are shaped by EU actors, such as Europeanized industry actors, the Commission and the European Parliament (hereafter: the Parliament). National fields will be characterized primarily by bottom-up preference formation: the positions of national governments are shaped by national industries and interest groups.

Multi-level Governance: Loose Structures with Ample Room for Entrepreneurship

Similar to LI, the MLG approach focuses on the policymaking processes proper, but it portrays the structural configuration of EU policymaking differently (see Hooghe and Marks 2001). Where Moravcsik pictures the distribution of authority as neat and clear, the MLG tends to portray it as rather messy. The core assumption is that political authority is dispersed across levels: 'authority and policymaking influence is shared across multiple levels of government – sub-national, national and supranational' (Marks, Hooghe and Blank 1996: 342). Because the EU organizations and member states share functional responsibilities in main issue-areas, they both exert influence over the development of EU policy (Eising 2004: 215). Decision-making arrangements among public and private actors abound, and the variation between sectors is considerable (Jachtenfuchs 2001: 25, Kohler-Kock and Rittberger 2006). Jurisdictions will partly overlap and operate on differing scales – rather akin to a marble cake (Marks and Hooghe 2004: 21). Here we do not find stable patterns of domination and subordination, but 'a wide range of public and private actors who collaborate and compete in shifting coalitions' (Marks and Hooghe 2004: 21).[1]

The Commission, the European Court and the Parliament all have independent influence over policy outcomes (Marks, Hooghe and Blank 1996: 346). Industries will mobilize directly in the European arena or use the EU as a public space to pressure state executives into particular actions (Marks, Hooghe and Blank 1996: 356). Moreover, interest groups and firms are believed to interact with governmental and parliamentarian organizations at national and European level in conjunction (Eising 2004: 212). Because the decision processes are loosely coupled, a policy debate at one level and in one arena does not necessarily trigger changes at another level or another arena (Eising 2004: 216).

The MLG perspective is not a coherent grand theory, but a looser description of policy shaping forces that draws on governance, network and rational-choice approaches (Marks, Hooghe and Blank 1996, Jachtenfuchs 2001, Bache and Flinders 2004, Kohler-Koch and Rittberger 2006). Critics have dismissed MLG as 'a descriptive metaphor rather than a theory' (Moravcsik and Schimmelfenning 2009: 68, see also Jordan 2001 and Fairbrass and Jordan 2004). Arguably, the weakest point in the MLG literature is that it lacks clear conceptualizations of the mechanisms that drive policy development. Yet, many central contributions attribute considerable weight to the power of extraordinarily skilled actors able to exploit the opportunities to further their interests (Hooghe 2001, Hooghe and Marks 2001, Eising 2004, Kohler-Kock and Rittberger 2006: 38). Moreover, scholars of governance and networking tend to portray EU policymaking in a very similar way as do MLG scholars (see Coen 1998, Peterson 2009: 106). Hence, we will draw on the contributions of these authors and others that highlight the importance of entrepreneurship in order to specify the causal mechanism of this perspective.

We can distinguish entrepreneurs from other actors by their political skills or the intensity with which they engage. Here, the reasoning of political scientist Robert Dahl remains valid. Back in 1961 he argued that whether actors are entrepreneurs or not depends on how they use their resources and also how '*skillful* or *efficient* they are in employing them' (Dahl 1961: 272 [italics in original]). Dahl specified this by stating that '[s]kill in politics is the ability to gain more influence than others, using the same resources' (Dahl 1961: 307). Also sociologist Neil Fligstein

---

1 Note also that Marks and Hooghe (2004) state that in some instances the patterns of power distribution in EU policy is rather clear and less opaque, but this runs counter to most contributions to this literature.

has argued that entrepreneurs are skilled societal actors, and that they will be 'more skilful in getting others to cooperate, manoeuvring around more powerful actors, and generally knowing how to build coalitions in political life' (Fligstein 2001: 107). Both scholars define entrepreneurs by their intrinsic qualities, not by their structural or institutional position. Further, their contributions indicate that creativity plays an important role, enabling the entrepreneurs to exploit cracks in the societal architecture and find new paths to influence.

On this backdrop, we define 'entrepreneurs' as actors that engage with extraordinary intensity and political skills. Policymakers and lobbyists who are merely performing their regular tasks will not be considered as entrepreneurs. Note that this means that we do not try to capture all aspects of agency under this perspective. Both organizations and persons can act as entrepreneurs (see Boasson 2011 for a more thorough discussion). Both Dahl (1961) and Fligstein (2001) argue that also *un*successful actors may be regarded as entrepreneurs – and so, to get a full view of the importance of entrepreneurship, we will need to explore unsuccessful as well as successful entrepreneurship. The MLG approach implies that the very character of EU policymaking makes entrepreneurs more influential than what can be seen in national policymaking. Yet, we have no reason to assume *a priori* that all entrepreneurs will be able to affect EU policymaking to the same degree, especially since they may encounter forces of varying strength.

The MLG literature pays considerable attention to how entrepreneurs circumvent, undermine or alter the formal distribution of authority and information distribution in order to change policy outcomes. Two techniques are highlighted as especially important: networking and agenda-setting. First, networking is a means to mobilize allies and induce cooperation among others (Leca, Battilana and Boxenbaum 2006, Hardy and Maguire 2008). Networking skills involve the ability to find ways 'to join actors or groups with widely different preferences and help reorder those preferences (…) The trick is to bring enough on board and keep a bandwagon going that will keep others coming' (Fligstein 2001: 114). Networking is a tool that may enable the entrepreneur to supply decision-makers with new kinds of information. In order to create large networks, entrepreneurs must possess significant flexibility, and even be willing to adjust their political project and shift their targets (Fligstein and Mara-Drita 1996, Fligstein 1997). Some MLG scholars describe networks as fairly stable (Eising and Kohler-Kock 1999, Mazey and Richardson 2006: 250, Peterson 2009: 105); others emphasize their fragile nature (Coen 1998, Hooghe 2001, Hooghe and Marks 2001, Meijerink and Huitema 2010). Here, we focus on the looser networks that are initiated by specific entrepreneurs. More stable networks will be understood as more stable authority structures.

Second, it is widely acknowledged that entrepreneurial agenda-setting may be important in articulating and introducing new policy ideas into the legislative process (see Kingdon [1984] 2011, Mintrom 1997, Finnemore and Sikkink 1998, Huitema and Meijerink 2010). Entrepreneurs may gain access to actual decision-making situations if they can include actors with formal authority in their networks

(Mintrom 1997: 760, Hardy and Maguire 2008). Kingdon ([1984] 2011) has argued that entrepreneurs will constantly be shopping around in search of decision possibilities where they can succeed in getting their policy ideas on the agenda, and will skilfully exploit any 'windows of opportunity'.

Summing up, the more skilfully and intensively an entrepreneur engages in networking and agenda-setting, the greater the likelihood of that actor being able to shape EU policy outcomes. Member states have the final say in most decision processes, but their preferences may be shaped by top-down entrepreneurship (for instance, member-state officials are influenced by EU officials or European corporations), as well as bottom-up entrepreneurs (such as national industries or national governmental officials). Hence, we may find top-down as well as bottom-up preference formation, depending on which actors are the best entrepreneurs.

Moving from Approaches to Mechanism Specification

Good inferences depend entirely on adequate conceptualizations of the phenomena under study (see Box-Steffensmeier, Brady and Collier 2008). Each of the three perspectives presents rather simplistic explanations of EU policy outcomes, but we believe that the actual policy explanations are complex. First, we see most relationships in EU policy development processes as being characterized by equifinality, implying that there may be multiple causal paths to the same outcome (Mahoney and Goertz 2006: 11). Second, we assume that policymaking is characterized by complex interdependencies: between the explanatory factors, or between the explanatory factors and the outcomes (Hall 2003: 383, Rueschemeyer 2003: 315). Third, we posit that policymaking will tend to be path-dependent: various different factors are at work over time, often in slow-moving processes (Pierson 2004, George and Bennett 2005: 212, Streeck and Thelen 2005).

This view of causality directs our attention towards mechanism-based explanations. Such explanations aim to render observed regularities intelligible by specifying in detail how they were brought about (Hedström 2008: 321). Mechanisms are the paths or features that connect causes and effects. Most mechanism-centred scholars agree that mechanisms are the basic feature 'governing the process' (Hedström 2008: 320). A mechanism-based explanation is 'abstract, realistic, and it explains specific political or social phenomena on the basis of explicitly formulated theories of action and interaction' (Hedström 2008: 333).

It makes little sense to start building mechanism-based explanations out of thin air (see Bourdieu 1992: 248). Many scholars have contributed to conceptualize parts of the features and mechanisms that may help to shape EU climate policy outcomes. The challenge is to systematize 'these bits of sometimes true theory' and develop new specifications that may be traced and tested through empirical research (Rueschemeyer 2003: 328–29).

We have seen that the three theory approaches represent significant analytical variation, and provide a good starting point for specifying policy-shaping mechanisms. It is particularly clear that the three perspectives are based on different assumptions as to preference formation and power distribution. First, the LI approach sees interest formation as a bottom-up process: national industries shape the interests of the national governments. NI opens up for both kinds of interest formation: issue-areas embedded in national organizational fields will be characterized by bottom-up preference formation, whereas issue-areas embedded in European organizational fields will be characterized by top-down preference formation. Also MLG opens up for both kinds of interest formation, arguing that skilled entrepreneurs, at the sub-national as well as the pan-European level, can shape the preferences of EU member states.

Second, the perspectives present different views on the relative power of the various actors. LI sees actors that are economically dominant nationally as the most powerful. According to the NI view, however, it is the actors' positions within organizational fields that are important: pan-European actors will be influential in Europeanized issue-areas (actors that have Brussels offices and/or operate in many EU member states), whereas sub-national and national actors will be influential if the issue-area is largely national in character. Finally, the MLG approach, as we interpret it, sees actors' entrepreneurial skills as the main determinant of influence.

More specific mechanisms are needed when it comes to how industry, policy interaction and EU-external influence may shape the policy outcomes. In the following we discuss issue internal mechanisms with a particular view to industry, before moving on to discuss policy interaction and EU-external influence.

Industry: Influencing through National Governments, Institutional Feedback or Entrepreneurship?

Knowledge Status

Some EU climate policy assessments highlight industry as a braking block in the EU climate policy development (see for instance Gullberg 2008). In particular, scholarly accounts of the failed efforts of establishing a carbon tax have highlighted the 'braking-block role' played by industry (Skjærseth 1994, Ikwue and Skea 1994). Other pieces of evidence point in different directions, for instance highlighting the role of big oil companies as emissions trading frontrunners and sources of inspiration for the subsequent development of EU policy (Christiansen and Wettestad 2003, Victor and House 2006).

In a recent discussion of the role of industry, Wyn Grant concludes that more research is needed in order to understand how and to what extent industry contributes to shape EU climate policy (Grant 2011). The main impression is that industry has not set the agenda for EU climate policymaking, but remained the 'elephant in the room'. A more general piece on business and politics states that '[b]usiness is a key political actor. It dwarfs other interest groups in terms of resources and political displacement' (Coen, Grant and Wilson 2010a: 2). They also note that the relation between business and government 'has long been a stepchild within the [political science] discipline', and that this line of research is 'undersupplied with theory', particularly true when it comes to EU studies (Coen, Grant and Wilson 2010b: 9, see also Schmitter 2010).

Some 70 per cent of EU interest groups are business-oriented, with industry providing the largest collective of such interest groups (Coen and Richardson 2009a: 6). The literature on EU lobbying argues that incentives to consult are especially strong at the EU level, 'as the degree of uncertainty about conditions in the member states is rather high', and the Commission has a procedural ambition of involvement 'or even co-option of interests' (Coen and Richardson 2009b: 339–40). EU lobbying studies primarily pay attention to actors that try to achieve influence by approaching the policy process in Brussels, but thus far few conclusions have emerged as to the effectiveness of this lobbying activity (Coen and Richardson 2009b: 343–44). In this book, we do not focus explicitly on lobbying, but explore a wide array of channels through which industry may affect EU policymaking. In the following we apply the three perspectives to specify various mechanisms through which industry may shape EU climate-policy outcomes.

Liberal Intergovernmentalism: National Governments as Instruments for Industry?

LI portrays industries as rational actors promoting pure economic interests. Here it draws especially on theories developed in the US institutional setting, applying profit-maximizing models of the firm to explain political action (see discussion in Coen, Grant and Wilson 2010b: 13). Moravcsik argues that industry actors will be able to specify which EU steering method and competence distribution serve to maximize (or satisfy) their economic interests in specific issue-areas (Moravcsik and Schimmelfenning 2009: 68). Furthermore, governments will weigh the influence of national industry by 'their size and intensity of gains and losses' (Moravcsik, 1998: 36). Member-state governments will take inputs from various national actors into account, but will attach most weight to the demands from the economically dominant ones. Hence, national governments can be expected to act as instruments for dominant national industry actors.

The LI view leads us to propose one industry-influence mechanism: *instrumental governments*. The basic assumption is that economically dominant national industries have shaped the positions of national governments. This implies that we can expect similarity in the EU policy preferences of dominant national industries and national governments, resulting from a one-way causal relationship: industry actors shape the positions of the government, not the other way around. Industry actors will have a strong indirect influence on the EU policy processes, but hardly have any direct impact. Moravcsik (2008) emphasizes that the influence of national industries will be highly issue-specific: an industry with considerable influence in one area may be weak in another, depending on its economic interests in various issue-areas. He implicitly assumes that all dominant industries have a close relationship with national governments, and he does not discuss whether the transnationalization and Europeanization of dominant European industries have affected their ability to influence national governments.

The case studies allow us to assess the relationship between the economic interests of dominant national industries, the preferences of national governments and EU policy outcomes. According to the LI perspective, we will expect that *the emergence and character of EU climate policy outcomes will be shaped by dominant member states that defend the interest of their economically powerful national industries.* Any differences in policy outcomes across cases will stem from variance in the dominant industries and their interests.

New Institutionalism: Industry Influence Depending on Organizational Field Dynamics?

In the NI perspective, it is the institutional logic(s) and authority distribution of organizational fields that will shape how industry perceives its interests (Fligstein 2008, Scott 2008). Over time, social processes within the field may alter the industry actors' views on EU steering methods as well as competence distribution in various issue-areas. Industry, national governments and EU actors that participate in the same fields are interdependent, developing their interests through mutual social processes within the organizational fields in which they are involved. Changes in the positions and functioning of one type of actor will also affect the functions and interest perceptions of other actors. Industry may be embedded in a range of fields: a corporation may take part in many national organizational fields, or both national fields and larger European organizational fields. The importance and role of industry actors may vary from field to field. On the other hand, corporations and governmental organizations involved in the same organizational fields will develop similar preferences.

Dominant interest perceptions and the distribution of power among national governments, industry and EU organizations hinge on the logic of the field and the historical role of national and EU steering respectively (Boasson 2011). With the notable exception of Fligstein (2008), few scholars to date have employed an NI perspective in specifically exploring the policy influence of industry. Applying NI to understand the impact of industry, we propose one mechanism: *institutional feedback.*

First, the steering method favoured by industry will depend on whether the logic of technology development or the market logic dominates their organizational field(s). As the institutional logic dominant within a given field is applied in decision-making it will gradually become stronger and more refined, and the leading industry actors within the field will prefer new EU policies to be aligned to this logic (Pierson 1996, Lawrence and Suddaby 2006, Scott 2008). Application of the logic in day-to-day decision-making as well as discursive use of the logic will contribute to its diffusion and specification.

Second, the distribution of competence can be shaped by institutional feedback: if the EU starts to produce more of the rules that govern a certain organizational field, industry will pay more and more attention to the EU level of policymaking, and the more attention the industry actors pay to that level, the stronger will centralization of control within the field become (Niemann 2006: 35, Fligstein 2008: 216). Moreover, industry will increase its geographical outreach and Europeanize its activities.

Over time, this will lead industry to become more and more positive to strong centralization of control in all issue-areas that regulate the functioning of the surrounding organizational fields, and Europeanized industry will also increase its ability to shape the policy outcomes. Neo-functionalists (see Niemann 2006) highlight how this may spur increasing centralization of control. Fligstein (2008) also draws attention to the reverse process: when national organizational fields dominate and no Europeanized field exists, industry will tend to prefer a low degree of centralization. It is reasonable to expect that industry will tend to rely on links to national governments if they take part primarily in national organizational fields, but target EU policy processes more directly if they take part in Europeanized organizational fields.

The case studies allow us to assess whether there is a relationship between the character of organizational fields and the role played by industry in EU policy formation. From an NI perspective, we would expect that *the emergence and character of EU climate policy outcomes will be shaped by the dynamics of the organizational fields involved in the policymaking.* Any differences in policy outcomes across cases will stem from variance in the institutional logics and Europeanization of the organizational fields in which the industry actors are embedded.

Multi-level Governance: Industry Influence Reliant on Ability to Perform Entrepreneurship?

So far, MLG scholars have concentrated more on the entrepreneurship performed by the Commission than industry actors (Hooghe 2001, Hooghe and Marks 2001). Yet, these scholars tend to portray industry and other non-state actors as important players that enhance their influence through entrepreneurship. In line with findings in the literature on EU lobbying, we will expect that the ability of industry actors to affect EU policy outcomes will hinge on their entrepreneurial skills, not their financial resources (see Coen and Richardson 2009b: 347).

Applying MLG to understand industry impact, we can extract one mechanism: *network entrepreneurship.* This implies that industry entrepreneurs may use persuasion and exploit negotiation dynamics to change the preferences of any given actor over time (Eising 2002: 86). In practice, this means that industry actors skilled at navigating in the multi-level landscapes, creating networks and agenda-setting opportunities, will be able to change the positions of the key actors in EU decision-making (Meijerink and Huitema 2010). Industry entrepreneurs will aim to shape the preferences of member states. However, it will also be important for industry to include Commission officials in its networks, not least because of the Commission's access to information from member states, its technically skilled staff and its formal agenda-setting role (Eising 2004: 218). Moreover, due to the veto-ability given to the European Parliament and its role in the co-decision procedure, members of the Parliament (hereafter: MEPs) are increasingly important as industry allies (Hooghe and Marks 2001: 8).

The case studies allow us to assess whether industry actors influence policy outcomes through entrepreneurship. On this backdrop, we expect that: *the emergence and character of EU climate policy outcomes will be shaped by industry actors that exert extraordinary skills or intensity in the policy process.* From this it follows that difference in policy outcomes across cases will stem from variance in the entrepreneurial activity of industry actors.

Policy Interaction: Operating through Functional, Bargained or Institutional Interaction Mechanisms?

Knowledge Status

The first literature on European integration, the neo-functionalist literature, argued that policy outcomes in one part of the European Community would affect other policy areas. In fact, this was presented as the very motor for greater European integration (see Haas [1958] 2004). However, the term 'spillover' and not 'interaction' was used to denote situations in which developments within one EU policy area affected another.

Empirical evidence from the past 40 years indicates that interaction patterns tend to be multi-directional and complex, rather than one-directional and simple (see Niemann and Schmitter 2006), so the determinism and automaticity of the initial spillover maxim has now been discarded (Niemann 2006: 27). International environmental politics scholars have paid increasing attention to studies of 'horizontal' institutional interaction, concluding that different EU environmental policies have significant impact on each other (cf. Oberthür and Gehring 2006). As to the various EU climate policies, they have developed in fairly parallel processes and have been eventually become known as 'integrated' (see European Commission 2008b and Chapter 3 in this book). However, they have also been developed by different actors and have distinct and different political histories. Recent administration studies show that, as a general pattern, there is rather little communication across sectors and policy-areas (Egeberg 2006, Trondal 2010). That makes it relevant for us to scrutinize more closely the extent to which, and how, the various climate policies have affected each other.

Spillover discussions in neo-functionalism have focused mainly on how increased centralization of control in one policy area will lead to centralization in others (Haas [1958] 2004, Niemann 2006). In this book, we will also discuss how the steering method of one issue-area may affect the steering method in another policy area. This has attracted more attention in the literature on environmental regime interaction than in the European integration literature (see Stokke 2001, Oberthür and Gehring 2006, Oberthür and Stokke 2011). Our discussion of EU policy interaction will draw on EU policy-specific contributions as well as contributions focused on regime interaction.

The regime interaction literature defines 'policy interaction' as something that occurs if one policy area affects the development or performance of another policy area (Oberthür and Gehring 2006: 26). Our main focus is on the first of the two indicated here: the policy development part, not the performance part. We posit that *interaction will occur if one EU policy directly or indirectly affects the outcome of another EU policy process*. How one policy affects the implementation of a policy is taken into account only if this feeds back to the EU-policy development processes and leads to changes in policy outcome. This definition of interaction implies that initiatives aimed at ensuring coordination between policy areas are not seen as interaction *per se;* rather, such initiatives may or may not result in policy interaction.

Some scholars use 'interaction' to denote how EU policy affects national climate policy or how the climate regime affects EU policy, but we reserve the term for EU policies exclusively. Thus far, EU climate policy interaction studies have employed the term in the two first-mentioned ways (see Sorrell 2003, Oberthür and Gehring 2006). The question of interaction between different EU climate policies has remained largely unexplored. The direction of interaction is also important: a policy that influences others will be characterized as a 'sender', while one that influences others will be characterized as a 'receiver'. We will explore the extent to which the four policy areas act as senders as well as receivers.

We will apply the three traditional European integration approaches to specify different policy interaction mechanisms. These approaches enable us to take into account interaction both in specific decision situations (see Sebenius 1983, 1984, 2009), and when it plays out as longer feedback processes, operating over considerable periods of time (Pierson 2004). As we will note, the LI approach pays particular attention to the former, but also opens up for more long-term functional interaction. NI highlights institutional feedback processes, operating over long periods of time, and MLG is rather silent on the issue of interaction.

Liberal Intergovernmentalism: Bargained and Functional Interaction?

Moravcsik does not explicitly discuss policy interaction in his contributions. Yet, his consistent focus on the economic interests of industries and intergovernmental bargaining situations points towards the inclusion of two kinds of policy-interaction mechanisms.

First, Moravcsik sees economic structures as the foundation for EU integration. This opens up for a specific kind of interaction: *functional interaction*, or what Haas ([1958] 2004) called 'functional-economic spillover'. Niemann and Schmitter

(2006: 49) explain the functional-economical spillover of Haas by an expirical example: 'the viability of integration in the coal and steel sectors would be undermined unless other related sectors such as transport policy followed suit, in order to ensure smooth movement of necessary raw materials'. Young (2002: 23) describes this as 'functional linkages'; Stokke (2001) uses the term 'utilitarian interaction'. Young defines a functional linkage as 'when the operation of one institution directly influences the effectiveness of another through some substantive connection of activities involved' (Young 2002: 23). Stokke (2001) describes how a policy outcome may alter the costs and benefits available in another policy area. All these scholars seem interested in the same phenomenon: how the introduction of one policy can create tangible, economic pressure for change in another policy. More specifically, such policy-induced economic changes will enhance the economic power basis of certain governments and hence improve their clout on EU policymaking in functionally related policy areas.

Second, governmental officials may initiate policy linkages in bargaining situations so as to enhance their bargaining position in issue-areas of specific national interest: *bargained interaction*. Moravcsik (1993, 1998) explicitly argues that national officials will try to link issues in order enhance their impact during the negotiation processes. He stresses how skilled negotiators can use information and negotiation techniques to link issues in ways that enhance their clout. Like other scholars of bargaining theory, Moravcsik views negotiators as rational actors with specific tactical skills (see Moravcsik 1993, 1998, Sebenius 1984, 2009). Negotiators are assumed to have clear preferences as to which issues they will prioritize and which they are willing to sacrifice in order to gain concessions in others. This kind of interaction has arguably an entrepreneurial flair: whether certain issues are linked depends not only on the economic interests of the various parties. Also important are their negotiating skills when it comes to discovering possibilities for policy interaction.

The case studies will allow us to assess whether the policy outcome will be affected by the two interaction mechanisms derived from the LI approach. First, this will lead us to expect functional interaction: *the emergence and character of EU climate policy outcomes will be shaped by functional pressures from related policies.* From this it follows that any differences in policy outcomes across cases will stem from variance in functional interaction. Second, we expect to find bargained interaction: *the emergence and character of EU climate policy outcomes will be shaped by actors that initiate and bargain linkages to other policy areas.* Hence, differences in policy outcomes across cases will stem from differences in bargained policy linkages.

New Institutionalism: Institutional Interaction?

The core of the NI interaction argument is that later policy decisions will be conditioned by earlier ones (Pierson 2004, Streeck and Thelen 2005). Hence, attention is drawn towards how 'old' EU climate policies may affect the competence distribution and steering method of 'new' policies. NI scholars like Paul Pierson (1996) and Neil Fligstein (2008) have emphasized that former EU policy decisions will shape actors' interest perceptions in new issue-areas. We extract one mechanism from the NI perspective: *institutional interaction*.

Neo-functionalists as well as New Institutionalists argue that actors will have bounded rationality and any interaction effects will tend to be unintentional (Simon [1947] 1997, Fligstein and Stone Sweet 2002: 1209). Haas (1970) maintains that EU decisions are often made with highly imperfect knowledge of the consequences and frequently under the pressure of deadlines. This makes it likely that issue-areas will affect each other in incremental or even accidental ways. Historical Institutionalism goes further, arguing that unintentional consequences can be found also when the actors have complete information, act transparently, and do not face major time constraints (Fioretos 2011: 380). Pierson (1996: 139, 140) argues that '[t]he more "tightly coupled" government politics are, the more likely it is that actions in one realm will have unanticipated effects on the other'. 'Tightly coupled' refers to policies that build upon each other: when one policy changes, it requires changes in others. Further, Pierson argues that this is particularly likely in EU policymaking, since the EU has to deal with many tightly coupled issues.

Haas ([1958] 2004) showed how a policy introduced in one sector could create processes that spill into other sectors, changing how the actors perceive their interests. Haas called this 'learning'. We prefer the more neutral term 'institutionalization', to avoid possible normative judgements of the effects of such interaction. According to neo-functionalists, the European Commission may have an important role in this kind of interaction. Lindberg, for instance, argues that the Commission's cultivation of contacts with national civil servants and national interest groups will gradually lead to its 'informal co-option' of the national elites of member states, to help realize its European objectives (Niemann 2006: 19, Lindberg 1963: 71). International environmental policy scholars like Oberthür and Gehring (2006: 22) and Stokke (2001: 10) discuss how knowledge, ideas or norms produced within one international regime may affect another regime. For instance, one international agreement may serve as a model case for another, or norms and institutional logics may diffuse from one policy area to another.

By 'institutional interaction' we therefore refer to incremental, partly unintended institutionalization processes whereby some degree of centralization and steering method is transferred from one policy area to another. On this backdrop we expect that *the emergence and character of EU climate policy outcomes will be shaped by the character of historically dominant policy areas*. In line with this, we expect that any differences in steering method and competence distribution across cases will stem from differences in historical roots and dynamics.

Multi-level Governance: Bargained Interaction?

MLG pays less explicit attention to interaction than the two other approaches. However, it can be argued that bargained interaction, with its weight on entrepreneurial skills in the negotiation process, is in line with the MLG perspective. We have presented this interaction mechanism under the Liberal Intergovernmentalism perspective, since Moravcsik explicitly discusses issue coupling in bargaining processes. Yet, there is one significant difference between the two perspectives in this respect: whereas Moravcsik expects member states to initiate bargained interaction, it is in line with the MLG perspective to expect EU officials, from the Commission and the Parliament, to be the most important initiators. The case studies allow us to examine which kinds of actors actually initiate bargained interaction.

External Environment: Operating through Coercive Pressure, Institutional Diffusion or International Entrepreneurship?

Knowledge Status

By 'external environment' we mean economic and social features that are rooted outside the geographical area of the European Community, outside the political processes in Brussels and the member-state level. Even in the early 1960s, European integration theory was criticized for not paying enough attention to external factors (see Niemann 2006: 22). Yet, subsequent theory contributions continued to regard the European Community as a 'closed system'. Studies of EU environmental policy have, however, begun to pay more attention to how the external environment, the climate regime in particular, affects EU policymaking (see Skjærseth and Wettestad 2002, Cass 2005, Oberthür 2006, Oberthür and Pallemaerts 2010, Oberthür and Dupont 2011). These studies show that the international developments and EU internal processes are linked – but we still know little about the specific mechanisms at work.

How can existing theory help us to understand how EU-external factors diffuse into EU policy processes? None of the three dominant theory approaches presented here pay much explicit attention to such external factors. However, the more overriding theoretical schools in which the three approaches are embedded may offer some guidance. The three approaches highlight different aspects of the EU-external environment: Liberal Intergovernmentalism directs attention towards external coercive pressure, whether economic or legal; New Institutionalism highlights softer, socially-driven diffusion processes; Multi-level Governance emphasizes the importance of entrepreneurs bringing the external impulses into EU policymaking.

Liberal Intergovernmentalism: Coercive Pressure

The LI perspective and its neo-realist roots give us reason to believe that changes in international market conditions, as well as binding international commitments in regimes with strong compliance systems, may force the EU to adopt policies with certain characteristics. Moravcsik argues that states are driven by issue-specific preference functions about how to manage globalization: member states will promote EU policy developments that protect their economy against the negative aspects of globalization and enable it reap profits from globalization (Moravcsik and Schimmelfenning 2009: 70). Further, Moravcsik posits that member states opt to yield some sovereignty to the EU if this is in line with their interests. It follows from this that also the European Community may yield some of its sovereignty to other international organizations with strong compliance regimes. On this backdrop, we derive a *coercive pressure* mechanism from this perspective.

First, changes in international market conditions represent an important source of changes in the economic interests of industries as well as member states. Member states are likely to opt for climate policy outcomes that can enhance their global economic competitiveness, and avoid options that may harm their competitive positions vis-à-vis non-EU states (see Knill 2005: 7). Thus we are unlikely to see the emergence of, for instance, EU climate policies that could hamper the global competitiveness of certain industries, or change the regulatory competition between EU member states and non-EU states. Moreover, EU member states may call for certain policies aimed at strengthening their situation in global markets. Second, the EU may succumb to international or supranational law (see Knill 2005: 7). This is particularly relevant for climate policy, since the EU is a party to the Kyoto Protocol. It can be assumed that the more formal competence that is transferred to the international regime and the stronger its compliance regime, the more likely will it be to shape EU-internal policy outcomes (see Stokke, Hovi and Ulfstein 2005).

This leads us to expect that *the emergence and character of EU climate policy outcomes will be shaped by international market conditions or international law.* Hence, any differences in policy outcomes across cases will stem from variance in international market conditions and international juridical obligations.

New Institutionalism: Institutional Diffusion?

NI approaches to European integration have not paid much attention to how factors outside the EU may play into EU policymaking. However, the wider sociological institutionalist community has shown interest in how the international environment may affect national policy processes (see Fioretos 2011: 375), so we can apply sociological institutional contributions to deduce a new EU-external institutional mechanism. Several NI scholars have studied how the emergence of global norms, standards, policy fashions and soft rules tend to diffuse globally (see Meyer et al. 1997, Meyer 2000, Drori, Meyer and Hwang 2006). They discuss how local institutions, policies and practices may be undermined, strengthened or changed as a result of international developments. This enables us to formulate one NI mechanism: *institutional diffusion.*

The diffusion of political norms, such as climate concerns and specific climate-policy remedies, can be enhanced by several factors. First, the EU may mimic successful policies in other countries and regions – the more actors that adopt a certain model, the stronger the pressure for the remaining organizations to adopt it (Meyer and Rowan 1991). The cumulative effect of many countries adopting the same norm, institutional logics or policy recipe will create 'peer pressure' among countries (Finnemore and Sikkink 1998: 903). Second, international organizations may promote specific policy recipes. Finnemore and Sikkink (1998: 900) argue that such actors can induce diffusion of norms by emulation (of heroes), praise (for behaviour that conforms to the group norms), or ridicule (of deviation). Third, the more a recipe diffuses, the more legitimized and theorized it will become, and the stronger will be the institutional pressure for adoption (Meyer et al. 1997, Meyer 2000). Theorization means self-aware development and specification of abstract categories and the formulation of patterned relationships, such as chains of cause and effect (Strang and Meyer 1993: 492). Institutionalization refers to the situation in which a certain policy is widely accepted, as both appropriate and necessary, and hence serves to legitimate the EU as a serious climate actor (Tolbert and Zucker 1983: 25).

This forms the basis for the expectation that *the emergence and character of EU climate policy outcomes will be shaped by norms and policy recipes adopted and promoted by other countries and international organizations.* Accordingly, any differences in policy outcomes across cases will result from variance in the external environment.

Multi-level Governance: International Entrepreneurship?

MLG scholars have generally paid little attention to EU-external features, but it is in line with the basic thinking to assume that actors can exploit global-level developments to promote their own interests. Both the architecture of international environmental agreements and the set-up of the EU are loose enough to provide room for entrepreneurial activity (Marks and Hooghe 2004: 24–25). MLG provides the basis for one specific mechanism: *international entrepreneurship.*

Just as entrepreneurs use EU-internal factors to create networks and exploit windows of opportunity, they may use developments in global arenas to this end. First, actors may develop networks that include global actors as well as those at the European and national levels. If actors find it advantageous to approach international climate negotiations in order to gain influence, they will do so (see Schattschneider 1960). Moreover, EU actors can create network contacts to states and international organizations outside the EU to obtain more information and credible technical inputs to the EU debate (see Mintrom 1997: 739–40). Second, EU actors may make strategic use of developments within the international climate regime in order to shape EU agenda-setting. Crucial moments in the international negotiations may create windows of opportunity, whereby the entrepreneurs have greater possibilities of link their 'pet issues' to actual decision situations (Cohen, March and Olsen 1972, Kingdon [1984] 2011).

Hence, we expect to find that *the emergence and character of EU climate policy outcomes will be shaped by entrepreneurs that use international networks and windows of opportunity strategically.* Any differences in policy outcomes across cases will result from variance in the international entrepreneurship performed in the each case.

Summarizing: The Nine Mechanisms

Table 2.2 sums up the nine mechanisms we have extracted from the three theory approaches.

Table 2.2   Industry, policy interaction and external environment mechanisms

Theory topics	Perspective		
	Liberal Intergovernmentalism	*New institutionalism*	*Multi-level governance*
Industry	Instrumental governments	Institutional feedback	Network entrepreneurship
Policy interaction	Functional interaction Bargained interaction	Institutional interaction	Bargained interaction
External environment	Coercive pressure	Institutional diffusion	International entrepreneurship

Method: Comparative Case-study Selection and Techniques

Comparative Case-study Selection

Comparative case-study methodology is particularly well suited for developing new typologies, synthesizing previous theory contributions, developing arguments concerning mechanisms and proposing more holistic theoretical frameworks. This made this method a natural choice for this book. George and Bennett (2005: 5) define the case-study approach as: 'the detailed examination of an aspect of a historical episode to develop or test historical explanations that may be generalizable to other events'. A 'case' is defined as an instance of 'a class of events', with an event being 'a phenomenon of scientific interest' (George and Bennett 2005: 17). As a minimum, every case study must specify the boundaries of the case and justify what it is 'a case of' (Rueschemeyer 2003: 320, George and Bennett 2005: 18). The method is 'structured' in that the researcher poses the same set of general questions for each case under study, thereby ensuring standardized data collection and enabling systematic comparison (Collier 1993: 105, George and Bennett 2005: 67). It is 'focused' in that it deals with only certain aspects of the historical cases examined (George and Bennett 2005: 67). Further, it is underpinned by careful case selection.

EU climate policy is the 'class of events' that is assessed in this book. We have selected our EU climate-policy cases on the basis of three criteria. First, we wished to deal with specific theoretical topics of interest: industry, policy interaction and the EU's external environment. All cases have been approached with the same research objectives in mind: to understand to what extent and how the policy outcomes are affected by industry, policy interaction and the EU-external environment. Having a consistent research strategy increases the likelihood of success in developing pertinent findings with regard to theory (George and Bennett 2005: 70).

Second, case selection is ideally guided by attention to where a specific theoretical phenomenon could be expected to be least or most likely to be detected (George and Bennett 2005: 121, Gerring 2008). In our case, we chose to focus on three topics of theoretical and practical interest, with no clear formulation of mechanisms at the outset. That meant we chose cases where we expected the three theory topics to have played some kind of role, whether an important or a non-important one.

We picked two cases where industry could be expected to have played an important role (the ETS and renewables) and two where the converse was to be expected (CCS and buildings). Further, we chose two cases where policy interaction could be expected to be highly significant (CCS and buildings) and two contrasting cases (the ETS and renewables). Finally, we expected the external environment to be crucial in two cases (the ETS and CCS) and largely irrelevant in the two others (renewables and buildings). Hence, the cases are matched with respect to the role of industry, interaction and external environment, and the three theory topics function as the focal points of the assessment (see Collier 1993: 112).

Third, there is a more pragmatic element involved as well. Both authors have previously carried out substantial work on several of the cases, exploring somewhat similar and different research questions (see Wettestad 2005, Skjærseth and Wettestad 2008, Boasson 2011). This has offered a good foundation for exploring the various mechanisms at work.

Application of Comparative Case-study Techniques

The book combines three comparative case-study techniques: theoretical pluralism, process tracing and pattern matching. *Theoretical pluralism* involves applying and combining different theories to explain a certain outcome. It takes seriously the fact that observation is theory-laden, and enables the researcher to offer a more appropriate explanation for each individual case (George and Bennett 2005: 111). We have applied and combined several theories in order to seek valid explanations for each of the four policy outcomes, and to develop new and, we hope, more precise conceptualizations of the mechanisms at work. We have cast the net wide and included the most-debated approaches to European integration as well as other approaches prominent in the social sciences. Not least, we believe we have combined sociological New Institutionalism and political science theories in a novel way.

Process tracing is a systematic within-case analysis that aims at identifying the actual associations between causes and outcomes (Hall 2003: 397, Mahoney 2003: 363). Researchers who apply process tracing often look at a finer level of detail than the proposed theoretical explanation, aiming to document that the processes within the case fit the theory (Bennett 2008: 705). Moreover, process tracing helps the researcher to sort out dependencies between the cases from the case-internal mechanisms at work (George and Bennett 2005: 33). This method involves combining a range of empirical sources – interview transcripts, archival documents, historical accounts and other sources (George and Bennett 2005: 86).

In order to achieve good process tracing, we have sought to map in detail the processes through which the various policy outcomes have come about. This involves both going back in time to trace the roots of the policies, and consulting a range of sources. In addition to scrutiny of relevant previous studies, this book builds upon a close study of EU, national and industry documents and of relevant news services like *ENDS* and *EU Energy*.

This has been further complemented by a substantial number of in-depth interviews with key Brussels actors – representatives of EU institutions, member states, environmental organizations, targeted industries, and academics/observers – and some selected Norwegian actors (see list of interviewees). These interviews were conducted in face-to-face meetings (one by Skype), in several rounds in the period 2008 to early 2012. Interviews were semi-structured: they addressed a core set of questions, but were tailored to the special experiences and competences of each interviewee. All interviews were conducted on the basis of the principle of anonymity. We feel that this helped the interviewees to speak much more freely and openly than they would have done otherwise. The drawback is of course that we cannot reveal the specific source of comments and statements. This is a general dilemma in such interviewing. What we have done is to indicate the rough placement of the various interviewees among the main groups – for instance, 'EU officials', 'civil society representatives'– which should provide the necessary background.

Due to the complexity of EU policymaking we have faced tough choices as regards member-state focus. This book covers policy development processes over a considerable length of time, and seeks to probe deeply into how EU member-state positions were shaped and transformed over time. In all cases we have tried to cover the broad picture of the preferences and actions of member states. In addition, we have selected a set of member states to follow in greater detail: Germany, Poland, Spain, Sweden, and the UK.

We think this set fulfils a range of important criteria. First, it covers main actors in the development of EU climate policy. These five countries are some of the largest member states with the most powerful positions in the EU. Moreover, these countries have played key roles in EU climate-policy development, whether as leaders or laggards. Second, we have included both northern and southern member states. Spain was chosen both because of its considerable size and its key role in some of the policy development processes under study, renewables in particular. Third, countries from the 'old EU'/'the EU 15' as well as the 'accession states'/'EU 12' are included. Poland is particularly interesting due to its leader position among the EU 12 and its importance in the development of the EU climate package, playing a key role with respect to the ETS, renewables and CCS. Fourth, since the Nordic countries have long maintained a comparatively green profile, Sweden is included in our group of selected countries. Sweden is a relatively small country, but has played a notable and special role in relation to several issues, renewables and energy policy for buildings in particular. Fifth, we have chosen to focus on countries that have a particularly interesting relationship to key industries related to the development of EU climate policy. Germany is notable in this respect, since it is home to both some of the most central emitters as well as some of the most prominent low-carbon industries, like wind and solar equipment producers. Also the UK stands out in terms of important industrial characteristics, with its oil industry and its special electricity market where there are no dominant British utilities.

In addition, we seek to cover other member states that have played key roles in the different stories. France, for instance, is certainly one of the dominant actors in the EU policy development game; moreover, it held the EU Presidency at a crucial moment – when the climate package was negotiated in 2008. We take into account this special role of France, but since it has generally had a modest role in the development of EU climate policy it was not included in our selected group.

Third, *pattern matching* involves systematic comparison of mechanism and typologies across carefully matched cases (Mahoney 2003, George and Bennett 2005: 180–204). Thus it is fairly close to the comparative historical work discussed by Rueschemeyer (2003), which uses within-case and cross-case analysis. Such analysis explores more complex interactions among causal factors, and traces and compares multiple paths of causation, without assuming a linear relation between the independent and dependent variables. In this book, we have first matched and confronted the mechanisms derived from theory with the actual cases and on this backdrop specified the formulation of the mechanisms. Then we have matched the mechanisms and causal stories in each of the cases, and on this backdrop developed new mechanisms and suggestions as to the relationship between certain outcomes and certain mechanisms. It is to be hoped that the latter venture has resulted in propositions that can help to explain how equifinality, complex interdependencies and path dependency may influence the development of EU climate policy.

Before turning to the four selected climate policies in turn, let us first summarize the overall development of EU climate policy.

Chapter 3

EU Climate Policy: From Pieces to Package

Introduction

This chapter gives a brief presentation of important actors, crucial decision situations and breaking points in the EU climate policy history, from the time when climate issues were first placed on the EU agenda in the 1980s and up until 2010. It provides an important factual backdrop for the in-depth case analyses and the subsequent comparative assessment to follow.

We distinguish three main phases in the development of EU climate policy: pre-1997 (up to and including the Kyoto Protocol); 1998 to 2004 (gearing up to the new drive); and 2005–2010 (the new drive and beyond). Two of the cases studied in depth – renewable energy and energy policy for buildings – have histories that pre-date the 1980s, but since the late 1980s they have been part of the EU climate policy portfolio. We focus primarily on the relationship between the development of EU climate policy and broader international developments, especially within the global climate regime, with special attention to the 'new drive' after 2005.

In addition to overviews of main policy developments in each of the three phases, we discuss the evolution of main roles and positions of important actors – the European Commission and the European Parliament, member states, industry and environmental organizations. The subsequent case chapters in the book are also structured in accordance with the three main phases, providing more detailed descriptions of how deliberations and collaborations involving the groups of actors have developed in our four focal issue-areas: emissions trading, renewables, carbon capture and storage (CCS), and energy policy for buildings.

This chapter sheds light on why EU climate policy emerged in the first place, and why there has been gradual shift towards increased centralization of control, and a combination of market steering with technology development steering. Three main types of empirical sources have been used. First, valuable insights have come from the literature on EU climate policy. Second, we sum up main elements of central EU policy documents, particularly from the Commission and the Parliament. Third, we draw upon our pool of interviews conducted with key actors in and observers of EU climate policymaking.

We begin with the period leading up to 1997 and the Kyoto Protocol, a phase characterized by rhetoric and somewhat symbolic policies. Then we move on to the 1998–2005 phase, when there was a radical shift in thinking on carbon pricing, but less development in other areas of climate policy. Finally, we discuss the new drive from 2005 and 2010, which has resulted in greater centralization and a broader portfolio and package of EU measures.

Pre-1997: A Shaky Start

Main Policy Developments

Concern about global climate change goes back to the 1950s, but up until the 1980s discussions were mainly of a scientific nature, and were more energetic in the USA than in Europe (see Jäger and O'Riordan 1996, Jordan and Rayner 2010). In October 1985, the United Nations Environment Programme (UNEP), the World Meteorological Organization (WMO) and the International Council for Science (ICSU) co-hosted a scientific meeting in in Villach, Austria. This resulted in the Villach Resolution, which contributed to place the climate issue on the EU agenda (see European Commission 1988).

It was in fact the European Parliament that adopted the first official EU document on the climate change in the form of a resolution in 1986, setting in motion the interplay between developments at the global stage and EU developments. This resolution acknowledged and recognized the complexity of the issue and requested other main EU bodies and actors to prepare calls for action and relevant policy measures designed to combat climate change (European Parliament 1986). Yet, climate change did not immediately become one of the EU's overriding environmental policy concerns: for instance, it was not mentioned as a priority issue in the EU's Fourth Environmental Action Programme (1987–1992) (European Commission 1987b).

International policy discussions intensified towards the end of the 1980s (Jäger and O'Riordan 1996). The Toronto Conference in June 1988 stands as a landmark for the heightened international attention to climate change. This Conference produced the first proposed international emissions reductions target: a 20 per cent reduction of CO_2 emissions by 2005 (with 1987 as the baseline) (see Andresen and Boasson 2012). In the same year, UNEP and WMO established the Intergovernmental Panel on Climate Change (IPCC).

This spurred further EU action. The first Commission Communication on 'the "greenhouse issue" was issued in 1988, recommending further scientific studies and review of policy options. The Communication called for 'urgent action to reinforce and expand efforts in the fields of energy savings, energy efficiency improvement, development of new energy sources, use of safe nuclear technology' (European Commission 1988: 53). The need for a global policy horizon was also emphasized: 'The greenhouse effect is a global problem, the Community should therefore play an important part in the definition of global policy, involving in particular developing countries, towards a sustainable development' (European Commission 1988: 52).

The interplay between international developments and EU initiatives continued. During the spring of 1990, the use of economic and fiscal instruments in environmental policy was discussed within the Commission, including the question of a tax on CO_2 emissions, hereafter referred to as a 'carbon tax'. The European Council had climate change on its agenda for the first time in June 1990, with a call for the early adoption of targets and strategies for limiting emissions of greenhouse gases (Haigh 1996). In October 1990, prior to an upcoming World Conference on Climate Change, Council agreement was reached on stabilizing EU CO_2 emissions at 1990 levels by 2000. This agreement was based on the assumption that other leading countries would agree to take on commitments along similar lines. The Commission then proceeded to flesh out a more specific array of policies: fiscal measures, particularly a carbon tax; measures targeting the transport sector, including the possibility of a maximum speed limit for private cars; measures to improve energy efficiency, including a revitalization of the existing SAVE energy efficiency programme; and the new ALTENER programme on renewables.

In October 1991, a new Commission Communication clearly expressed the EU ambition to act as a global leader at the June 1992 United Nations Conference on Environment and Development in Rio de Janeiro (European Commission 1991). Later the Commission outlined four main measures for the Council to adopt: a framework directive on energy efficiency (within the existing SAVE Programme); a decision concerning promotion of renewable energies (the ALTENER Programme); a directive on a combined carbon and energy tax, on the condition that such a tax was also adopted by 'main competitors' within the OECD; and a decision concerning a monitoring mechanism for CO_2 and other greenhouse gas emissions (European Commission 1992b).

Already at this stage, the Commission was linking various climate policy issues. For instance, the 1991 Communication mentioned 'a package of measures' (European Commission 1991: 4). Moreover, a burden-sharing proposal was developed, where certain countries with higher development needs were to be accorded greater flexibility than the others (Ringius 1999: 139). The Commission argued strongly for the introduction of a carbon tax, for instance stating:

> It seems difficult to motivate economic agents to improving their energy efficiency if energy prices are too low. Moreover, some energy sources (certain renewables in particular) which are favourable for the stabilization objective as well as for overall environmental quality will not be able to develop significantly unless their market position can be enhanced by the internalization of their comparative environmental advantage into their price. (European Commission 1992b: 7)

However, the Council did not manage to adopt any of the proposals prior to the Rio Conference. This was mainly due to the carbon tax proposal, which proved too controversial for member states to accept. In particular, a small but active group of member states led by the UK campaigned against it. Industry was also fiercely opposed. All four of the proposed measures were based on Article 130 of the Single European Act, which meant that they all had to be adopted unanimously. Hence, the proposed carbon tax was not adopted and the EU's bid for international leadership began to unravel (Macrory and Hession 1996, Haigh 1996). This situation left the EU delegation to the 1992 Rio Conference with ambitious targets, but no convincing internal policies and measures for achieving them (Skjærseth 1994). Environment Commissioner Carlo Ripa di Meana resigned from his post in protest.

Despite internal differences prior to the Rio meeting, the EU pushed for binding commitments for industrialized countries in the international negotiations, seeking to get them to commit to stabilize CO_2 emissions at 1990 levels by the year 2000. This was not met with much enthusiasm from other OECD countries, but the United Nations Framework Convention on Climate Change (UNFCCC) was still adopted – and a main element in this convention was the non-binding commitment for industrialized countries, 'individually or jointly', to stabilize their CO_2 emissions by 2000 at 1990 levels (Jäger and O'Riordan 1996, Rowbotham 1996).

After Rio, the Commission gave priority to getting the climate policy package adopted. Getting the monitoring mechanism in place proved least difficult, and this was adopted in 1993. Related to the generally increasing 'subsidiarity wind' in the EU in the early 1990s, with opposition to EU-level policies, the budgets for both SAVE and ALTENER were considerably reduced, and watered-down versions of both programmes were adopted in September 1993 (Collier 1997, Jordan and Rayner 2010). Getting the carbon tax adopted continued to be problematic. Hopes of consensus were dwindling, and there was no sign of the main OECD competitors establishing carbon taxes. So the idea of a common tax was downplayed towards the end of 1994. Instead, the European Council called for a consideration of common parameters to enable member states to apply a CO_2 tax (Skjærseth 1994, Wettestad 2000). A first proposal to limit CO_2 emissions from passenger cars was tabled at an Environment Council meeting in 1994, setting in motion a discussion in subsequent years on how best to tackle such emissions (ten Brink 2010).

In the preparatory process for the first UNFCCC Conference of the Parties (COP) to be held in Berlin in 1995, there were again diverging positions within the EU. Still, the Council managed to reach agreement that the EU should aim for the launch of negotiations on a new protocol with strengthened commitments, to be concluded by the time of the third COP in 1997. This was included in the negotiation outcome, the 'Berlin Mandate' (Jäger and O'Riordan 1996). As to the establishment of overriding objectives for EU climate policy, we may note that the EU first adopted the objective that 'global average temperatures should not exceed two degrees above pre-industrial level' in 1996 (Oberthür and Pallemaerts 2010: 33).

The final deadline for tabling proposals for a protocol under the UNFCCC was March 1997. Early that month, after two years of internal deliberations, the EU was able to hammer out an agreement on a 15 per cent common target, combined with differentiated internal targets (Ringius 1999). This enabled the EU to stand out as the most ambitious of the major global actors. In other words, although it was still struggling with its internal policymaking, the EU had started to play an important role on the global stage (see Grubb and Gupta 2000, Ringius and Gupta 2001, Schreurs and Tiberghien 2010, Oberthür and Dupont 2011).

In Kyoto in December 1997, the EU spent considerable time on internal coordination and trying to stand up against the US desire for a protocol based on flexible mechanisms such as emissions trading (Grubb, Vrolijk and Brack 1999, Oberthür and Ott 1999, Andresen and Boasson 2012). It has been argued that without the effort from the EU, the deal would have been significantly weaker (see Grubb, Vrolijk and Brack 1999). Although the EU succeeded in getting a protocol adopted with strengthened commitments and with the USA on board, the more specific design of the Protocol had a distinctly American flavour (Cass 2005). Three flexible mechanisms became central ingredients of the Kyoto Protocol: emissions trading, Joint Implementation (JI), and a Clean Development Mechanism (CDM). The EU's own target became an eight per cent reduction of greenhouse gases (GHGs) by 2008–12 from 1990 levels. Furthermore, similar to other parties to the Protocol, the EU should make demonstrable progress in achieving its commitment by 2005.

We may conclude that by 1997, climate policy had emerged as a distinct policy area in the EU – but the few policies involved were of little substance, gave the EU limited centralized control and were partly pre-existing energy policies now dressed up as climate policies. Although the package concept was introduced, only separate, piecemeal policies were actually adopted.

Actors, Roles and Positions

Back in the 1980s, the then 12 EU member states had only limited understanding of climate change, and thus gave scant priority to the issue. From the late 1980s, public concern about the seriousness of climate change was growing, but waned again towards the mid-1990s (Wagner 1997). Here, it is important to recall that at this point, other environmental issues still stood out as more pressing for many governments – for instance, the depletion of the ozone layer, marine pollution and acid rain (Andresen, Boasson and Hønneland 2012).

Furthermore, the sheer comprehensiveness and complexity of the climate change issue made it complicated for governments to address. In Germany, for instance, 'the global warming question seemed to fall between the institutional cracks of government during much of the 1980s' (Hatch 2007: 44). Powerful coal-industry interests and controversy over nuclear power contributed to the impasse there. Still, the adoption in 1990 of the target to cut its CO_2 emissions by 25 per cent by 2005 made Germany the top EU climate frontrunner. But it soon became clear that other key countries, among them France and not least the UK, were far less ambitious and in fact did not support a strong EU climate policy at all. These countries gave higher priority to other concerns, such as subsidiarity and liberalization of the energy sector (see O'Riordan and Rowbotham 1996, Cass 2007, Rayner and Jordan 2011). As regards the overall distribution of member-state positions, the 1995 enlargement brought in states with a generally 'green' profile – Austria, Finland and Sweden. However, these countries, and not least

Sweden, initially chose to campaign for other environmental issues – acid rain in particular (Wettestad 2002).

Turning to industry, we note that the comprehensive nature of the climate change issue makes it natural for a range of relevant industries to be involved – including various energy-producing industries, energy consumers such as the steel and cement industries, car manufacturers, and engineers. In the formative years of the debate on climate change policy, most industries showed scant concern for climate change. Oil majors like Exxon, BP and Shell were initially very negative (Skjærseth and Skodvin 2009), and renewables industries were almost non-existent (see Chapter 5). The carbon tax proposal led to some of the fiercest lobby activity ever seen in Brussels (Ikwue and Skea 1994, Skjærseth 1994).

Environmental organizations were increasingly concerned about climate change, but other issue-areas such as acid rain still had prominence. Here we should recall that the establishment of such organizations at the European level and their related presence in Brussels was in an early stage. Whereas the European Environmental Bureau had been in Brussels since 1974, a range of new organizations set up Brussels offices between 1986 and 1990, among them Friends of the Earth, Greenpeace and World Wildlife Fund (WWF) (Duwe 2001, Wurzel and Connelly 2011c). In addition, the European Branch of the Climate Action Network (CAN) was established: an international network of environmental organizations engaged in climate issues. The EU carbon tax was the main priority for the environmental organizations, but they also stepped up their involvement in renewable energy and energy efficiency issues. In addition, they stood forth with heavy normative criticism of flexibility mechanisms, including emissions trading (Skjærseth and Wettestad 2008).

Within the Commission, the environment directorate (DG Environment) was seen as a weak body, with a very small staff compared to other DGs. Some even claimed it was dominated by 'ecological freaks' (Grant, Matthews and Newell 2000: 21). Furthermore, there was no special unit dedicated to the issue of climate change. The administrative capacity and strength initially available to the Commission for underpinning its climate change positions and leadership aspirations was clearly limited. In addition came internal disagreements within the Commission, perhaps most notably between Environment Commissioner Ripa di Meana and Taxation Commissioner Schrivener in the carbon tax debacle of the early 1990s (Barnes 2011).

As to the European Parliament, we have already noted that it was a frontrunner in drawing the attention of the EU to the issue of climate change. With regard to the Parliament's more general position in decision-making, the 1987 Single European Act (SEA) changed its standing somewhat. The SEA introduced a 'cooperation procedure', whereby the Parliament gained the right to a second reading (a new round of discussions and decisions) for certain laws being considered by the Council of Ministers, notably those relating to aspects of economic and monetary policy. This development was taken further by the 1993 Maastricht Treaty, which introduced the co-decision procedure, with the possibility of three rounds or 'readings' on selected laws relating to the single market and the environment, and the possibility of a final round in a Conciliation Committee with equal Council and Parliament representation (see McCormick 1999). Although the Parliament managed to use its growing power to influence important processes such as the Auto-Oil process in the context of air pollution, the impression as regards climate change is that it managed only to exert what Burns and Carter (2011: 60) call 'symbolic leadership', with limited influence on EU politics and decision-making.

On this backdrop, we conclude that – apart from the efforts of the Commission, backed to a certain extent by the Parliament, environmental organizations and a few member states like Germany and the Netherlands – there was essentially little interest in developing an effective and varied EU-level climate policy in this period. It was agreed that the EU should act as a leader in the international negotiations, but there were significant conflicts over the internal climate strategy.

1998–2004: Getting into Gear

Main Policy Developments

In the 1997 Kyoto Protocol the European Community took on a commitment to reduce specified GHG emissions by eight per cent by 2008–12. However, this outcome was less ambitious than the EU's initial position, as set out in the spring 1997 burden-sharing agreement. Hence, the internal EU burden-sharing agreement reached before the Kyoto summit had to be revised. Due to domestic political changes, Germany and the UK now opened up for more ambitious targets than those they had initially agreed to (Jordan and Rayner 2010).

The new EU burden-sharing agreement reached in 1998 required Germany to reduce its emissions by as much as 21 per cent and the UK by 12.5 per cent. In contrast, Sweden was allowed to increase its emissions by 4 per cent and Spain by as much as 15 per cent (Wettestad 2001). Thanks not least to emissions reductions in the UK and Germany, the EU was already well on its way to achieving the stabilization target by 2000 – without any effective *common* climate policy in place. The Kyoto target created a need for more EU climate policy, but gave little guidance as to which measures to adopt (Jordan and Rayner 2010: 65). A main reference point for the ensuing EU climate policy discussions became how the EU could put in place policies to ensure compliance with the Kyoto Protocol. The Commission put forward several proposals for revised or new internal policies, and in 1998 it announced a reversal in its position on emissions trading (Skjærseth and Wettestad 2008). In the field of reducing car emissions, it was decided to go for voluntary agreements as the initial way forward. The first agreement was struck between the Commission and European car manufacturers in 1998, followed by similar agreements with Japanese and Korean manufacturers in 1999 (ten Brink 2010).

In March 2000 the Commission announced a new, multi-track climate policy approach (Wettestad 2001, Jordan and Rayner 2010: 67). The central element in the first track was the establishment of the first European Climate Change Programme (ECCP I), a multi-stakeholder programme involving actors from industry, member states and NGOs as well as independent experts (European Commission 2000a). The aim was to develop proposals for climate measures and policies, and to identify cost-effective ways for the EU to meet its Kyoto commitments. The Programme was divided into six working groups – four focusing on policies and measures in the energy consumption, energy supply, transport and industry sectors, and the other two on flexible mechanisms and research. This resulted in a comprehensive report, published in July 2001, containing various proposals for further policies and measures (European Commission 2001a).

The second track was the establishment of an internal emissions trading system (ETS) (European Commission 2000b). The Commission's ETS proposal was put forward in October 2001 (European Commission 2001c). It proposed the establishment of a fundamentally decentralized system for the pilot phase of emissions trading (2005 to 2007) and the Kyoto Protocol commitment phase (2008 to 2012). Member states were to be given clear authority to determine the total cap on allowances to be handed out to participating industries, the distribution of allowances among sectors, and actors/installations within industry (power producers as main targets, but also including several energy-intensive industries), and various other allocation matters (Christiansen and Wettestad 2003, see also Chapter 4 in this book). A broader policy development continued along the lines indicated by the ECCP. Yet, as we shall see, there can be little doubt that quick development of an ETS became the flagship of EU climate policy.

DG Energy administered a third track, developing renewable energy and energy efficiency policies. The Commission launched high ambitions for stronger common EU policies in both issue-areas, but policy development took longer than expected. Energy efficiency proved particularly difficult, with the development of a whole string of regulations, covering energy labelling of various products, energy policy for buildings and a general directive for energy end use-efficiency.

Parties to the Kyoto Protocol were required to show 'demonstrable progress' by 2005 (Slingenberg 2006). Within the global regime, much discussion centred on further fleshing out the institutional architecture of the flexible mechanisms, and, *inter alia,* a more complete compliance regime. In this period, the EU had to face a range of challenges that tested its determination to take the lead in combating climate change. The first challenge was the Conference in The Hague in November 2000, where the US government put forward the demand that credits for carbon sinks should be counted as emissions reductions. This was strongly opposed by some EU states, which again led to collapse in the talks.

In March 2001 the EU found itself confronted with a far greater challenge when President George W. Bush pulled the USA out of the still not ratified Kyoto Protocol. On the other hand, the US exit also opened up a window of opportunity for the EU to exert global climate policy leadership, and at the Environment Council in Gothenburg in June 2001 the ministers used this opportunity and decided to adhere to the Protocol ratification process. The 2001 Swedish Presidency had managed to secure a common statement from all member states, criticizing the US government as well as insisting on the importance of ratifying the Protocol also without the USA on board (Slingenberg 2006). A major diplomatic campaign was set in motion. This helped pave the way for reaching agreement on the rules specifying the Protocol and its follow-up in the Marrakesh Accords at the climate regime conference in November 2001. Further, it aimed at securing Russian ratification, as the Russian Federation alone stood for 17 per cent of emissions (Jordan and Rayner 2010, Oberthür and Pallemaerts 2010, Wurzel and Connelly 2011b). After the USA had left the Kyoto Protocol, Russian ratification was needed for the Protocol to enter into force.

A range of important EU decisions followed. In 2001, the EU adopted a directive on renewable electricity (European Parliament and Council 2001). This came about after a heated debate over whether to opt for market or technology steering, resulting in a compromise directive with rather low centralization of control. In the following year, the EU burden-sharing agreement became legally binding (Decision 2002/358/EC), the EU ratified the Kyoto Protocol, and a Performance of Buildings Directive was adopted (European Parliament and Council 2002).

The ETS Directive that was formally adopted in 2003 was largely in line with the Commission's proposal (European Parliament and Council 2003). That meant that the EU had not only abandoned its initial opposition to this market-based policy instrument, it had also become an emissions trading pioneer with the establishment of the world's first international emissions trading system with companies as main actors (Christiansen and Wettestad 2003, Wettestad 2005, Skjærseth and Wettestad 2008). A subsequent decision-making process produced a directive specifying the links between the EU ETS and the flexibility mechanisms under the Kyoto Protocol. Political agreement on the Linking Directive was reached in April 2004 (Flåm 2009). And then, in November 2004, Russia ratified the Kyoto Protocol, paving the way for entry into force of the Protocol.

To sum up, the EU broadened its climate policy portfolio in this phase, most notably by managing to establish a key market measure: the ETS. This was accompanied by other EU-level policies such as the directives on renewables and on the energy performance of buildings. Even though EU climate policy became far more elaborate during this period, it was still marked by low centralization of control, with significant authority left in the hands of the member states. Let us now see how key actors changed their climate involvement and their positions during this period.

Actors, Roles and Positions

Among the member states, many of the old coalitions and divisions remained. The proactive group expanded slightly, particularly with Sweden joining this camp. German leadership continued, now under the slogan of 'ecological modernization' (Jänicke 2011). In 1998, a Red–Green coalition came into power in Germany, with the Green Party taking over the Ministry of Environment. In 1997, the British

Labour Party led by Tony Blair had won the elections in the UK, bringing a more activist approach to climate politics (Darkin 2006).

Here we should note that British and German emissions had already been reduced, due to non-climate policy related processes. The 'dash-to-gas' from coal in the UK did not have a primary environmental policy motivation, and therefore brought down emissions without specific environmental policy abatement costs (Wettestad and Hals Butenschøn 2000). To some extent this was also the case with Germany, where re-unification and the related closures and restructuring of East German industry brought a decrease in emissions (Hasselmeier and Wettestad 2000).

The southern EU members remained generally sceptical, as exemplified by the conservative Aznar government in Spain (Costa 2011). The first eastward enlargement took place in 2004, with 10 countries joining the EU, but in the early years of the 2000s these had only observer status and did not influence processes much – apart from requiring a substantial amount of time and effort from the Commission.

The Commission sought to regain its policy leadership position not least through the Climate Change Programme, with the ETS as the new policy cornerstone. There was collaboration with Commission President Prodi and Environment Commissioner Wallström on a proactive role for the Commission in climate policy. When José Manuel Barroso became head of the Commission in 2004 he did not initially signal any specific emphasis on climate change or the initiation of a 'new drive' (Barnes 2011: 49). On the other hand, in these years the Commission was involved in several ambitious policy processes that probably directed attention and capacity away from climate change; prominent among these were EU enlargement and a new treaty for the EU, the Lisbon Treaty. The processes were related, as the latter was intended to adapt EU institutions to the effects of the former (Benson and Jordan 2010).

It seems as if the US withdrawal from Kyoto and the related turn of global events served to dampen EU-internal climate criticism and activism from the European Parliament. Its formal decision-making role and weight increased with the entry into force of the Amsterdam Treaty. This meant a further extension and simplification of the procedure of co-decision. But in key processes – notably, the ETS decision-making process – the Parliament failed to make a marked imprint on the outcome, although it actively sought to introduce more auctioning and market streamlining and centralized control than proposed by the Commission.

Generally, industry stepped up its involvement in EU climate-policy development in this phase. It has been claimed that industry actors overall responded to Commission proposals and did not proactively launch their own proposals (Grant 2011). These actors were generally more positive towards the Commission's proposals for market measures than technology development measures. Big oil companies were frontrunners in this field, with the BP volte-face in 1997 and the subsequent development of internal emissions trading programmes in both BP and Shell (Victor and House 2006). The electricity industry supported emissions trading and championed market measures for renewable energy. However, the smaller renewable energy industry favoured technology development steering instead of market steering in the latter case. Moreover, energy-intensive industries were far less enthusiastic than other industries in their support for emissions trading. With the character of climate policy becoming more complex and varied, the roles and positions of industry became more multi-faceted and heterogeneous.

With regard to the environmental organizations, we see a radical increase in their climate-policy engagement in this period (Wurzel and Connelly 2011c). They continued to support a wide array of climate measures, but they also became more positive to the cornerstone ETS instrument. Moreover, they coordinated their campaigns and developed an internal turf-sharing, so that they were able to follow the whole range of climate policy development processes. Furthermore, the Bush presidency and the US withdrawal from Kyoto contributed to a 'rallying around EU policy' effect, affecting the role and room for the activity of such organizations in a somewhat similar way as in the case of the Parliament.

In essence, this phase was characterized by efforts directed at Kyoto Protocol implementation, and the group of actors positive to a stronger EU climate policy expanded. Central industrial actors welcomed the adoption of the ETS as the cornerstone of climate policy. The US withdrawal from Kyoto in 2001 further strengthened EU internal unity in this issue-area. But entering uncharted regulatory landscapes as to EU policy, member states were not yet ready to hand over control to EU-level bodies, as seen for instance in the decentralized set-up of the ETS. Furthermore, while we can note increasing agreement to regulate CO_2 emissions through market steering, disagreement increased as to whether to opt for market solutions in relation to renewable energy.

2005–2010: The New Drive

Main Policy Developments

From the mid-2000s and onwards, the EU took up the challenge of securing compliance with the Kyoto Protocol, set to expire in 2012, as well as of negotiating an international successor agreement. A report on 'greenhouse gas emission trends and projections in Europe' issued by the European Environment Agency in 2004 concluded that 'the latest projections for 2010 show that neither existing domestic policies and measures by Member States to reduce emissions, nor planned additional domestic policies and measures, will be sufficient for the EU-15 to reach its Kyoto target' (European Environment Agency 2004: 3). Member states started to launch medium- and long-term emissions reduction targets, and the spring European Council meeting in 2004 concluded that it looked 'forward to considering medium and longer term emission reduction strategies, including targets, at the 2005 Spring Council' (European Council 2004: 20).

So in February 2005, the Commission responded by issuing the communication 'Winning the Battle Against Global Climate Change', laying out the international negotiation strategy of the EU (European Commission 2005c: 11). The objective was to establish a multilateral climate change regime post-2012, with meaningful participation of all developed countries and the participation of developing countries, to limit the global temperature increase to 2°C. No other countries had showed willingness to commit to specific reduction targets, and hence the Commission did not recommend the adoption of a specific EU target at that stage. However, it did note that the EU would have to strengthen its internal climate policy, drawing attention to energy efficiency, renewable energy, the transport sector, and carbon capture and storage (CCS) in particular. Furthermore, a second European Climate Change Programme (ECCP II) was announced.

Here we should note that the 2005 Spring Council meeting upped the ante in relation to the Commission, by emphasizing that it was 'the EU's determination to reinvigorate the international negotiations by (...) developing a medium and long-term EU strategy to combat climate change, consistent with meeting the 2°C objective. (...) the EU looks forward to exploring with other parties strategies for achieving necessary emission reductions and believes that, in this context, reduction pathways for the group of developed countries in the order of 15–30% by 2020 (...) should be considered' (European Council 2005: 15–16). This gave the Commission backing for the development of an ambitious climate strategy. A central EU official interviewee states, 'Let's be frank: Council decisions are really of very little importance, but in this case it actually represented something important.'

In 2006, the IPPC prepared its fourth assessment report, and finalization of the sub-report attracted substantial media attention (Andresen and Boasson 2012). In addition, the British government commissioned a report on the effects of global warming on the global economy headed by economist Nicholas Stern, showing net benefits to be gained from climate protection (Stern 2006). The Stern report profoundly affected the European climate debate (Schreurs and Tiberghien 2010: 87). This view is supported by many of our Brussels interviewees, who have held that it contributed to change dominant sentiments about climate change policies.

The Commission responded by insistent calls for climate action. As Oberthür and Pallemaerts (2010: 46) note, 'progressive climate policies received additional support because they could serve to re-legitimize European integration more broadly and provide a suitable area to strengthen the EU's role on the world scene'. A central EU Commission official interviewee emphasized that 'it is important to understand that the "new drive" did not result from pressure from any specific member state (...). Rather, the Commission was responding to what was in the wind.'

A growing focus was also developing on energy security, related not least to the EU's energy policy relationship with Russia. The eastern enlargements of the EU in 2004 and 2006 had brought in a group of countries highly dependent upon Russian energy supply and hence interested in all measures that might contribute to reduce their import dependence (Claes and Frisvold 2009). When, in early 2006, Russian Gazprom held back gas supplies on the pipeline to Europe through Ukraine, this caused loss of supply also to some EU countries, reinforcing existing mistrust and fuelling the Commission's attention to energy security (Wettestad, Eikeland and Nilsson 2012).

In March 2006, a Commission Green Paper on Energy was issued (European Commission 2006c). Here, open and competitive energy markets were emphasized, with positive environmental implications, 'as companies react to competition by closing energy inefficient plants' (European Commission 2006c: 5). Three main elements were emphasized within 'an integrated approach to tackling climate change' – energy efficiency, renewables and CCS. A 'new energy policy for Europe' was further discussed and laid out at the March 2006 Energy Council meeting. As to 'sustainable energy', ministers called for an action plan on energy efficiency and a long-term strategy for renewables (European Council 2006, *Euractiv* 2007a). The Commission followed up with an Energy Efficiency Action Plan that aimed at reducing consumption by 20 per cent by 2020 (European Commission 2006d).

Normally, the introduction of new EU policy tends follow a somewhat cumbersome procedure, with the Commission with first issuing a green book, then a white book, and finally more substantive policy proposals (Europa 2012). This time, however, the Commission opted for another route and decided to launch a package of measures more quickly. In January 2007, it published a set of important documents (European Commission a, e, f). Included here was a core vision on climate change policy (European Commission 2007a). With this proposal the Commission called for a range of actions and policies to strengthen climate policy and to support an independent target of 20 per cent reduction in GHG emissions by 2020. This figure would be raised to 30 per cent if other countries agreed to a global post-2012 agreement (European Commission 2007a: 5). The Commission hereby tried to bring EU policy in line with the results and recommendations of the IPCC and the Stern report (Jordan and Rayner 2010). Reflecting on why the Commission deviated from standard procedure, a central Commission interviewee stated, 'we did it simply because we decided to! To use a cliché, we had a so-called window of opportunity and we seized it.'

DG Environment and DG Energy worked in tandem, with the latter directorate being responsible for significant climate policy measures. Highlighting the link between energy and climate policy, the Commission identified in its 'Energy Policy for Europe' communication a range of other potentially far-reaching sub-targets, including renewables, repeating a call first put forward in 2006 for a binding increase from 7 to 20 per cent renewable energy in the EU's energy mix by 2020, and also a 10 per cent biofuels increase (European Commission 2007e). As to carbon capture and storage, the Council was called upon to provide a 'clear perspective' on when plants would need to install CCS and the establishment of a mechanism to stimulate the construction and operation of up to 12 large demonstration plants by 2015. With regard to energy efficiency, the Council was invited to endorse the target of reducing EU energy consumption by 20 per cent by 2020, as presented in the Commission's Energy Efficiency Action Plan. Furthermore, a specific Strategic Energy Technology (SET) plan was proposed, setting targets to develop a range of relevant technologies, backed up by existing and new spending on energy research. The vision and rhetoric were certainly bold: the EU aimed at gaining 'world leadership in a diverse portfolio of clean, efficient and low-emission energy technologies (European Commission 2007e: 15).

The Commission proposals were then followed up and unanimously endorsed by the European Council in March 2007, under the leadership of the German EU Presidency and Chancellor Angela Merkel. It was agreed to support the Commission's proposal for a 20 per cent reduction in CO_2 emissions by 2020, increasing to 30 per cent if other OECD countries followed suit. The meeting also established an EU-wide 20 per cent target for the use of renewable energies, and a non-binding 20 per cent improvement in energy efficiency. The seminal '20-20-20' targets were thereby adopted.

As to CCS, the Council requested the Commission to establish a legal framework for CO_2 storage, and endorsed the Commission's suggestion to draw up a policy to 'stimulate construction and operation by 2015 of up to 12 demonstration plants'. Furthermore, the SET Plan idea was endorsed and a road map for developing 'European Industrial Initiatives' for key 'low carbon emitting technologies' was called for. The latter placed CCS on the top of a list of six technologies (European Council 2007, Claes and Frisvold 2009: 219). The ministers invited the Commission 'to submit the proposals requested in the Action Plan as speedily as possible' (European Council 2007: 14).

How then did these targets come about? The Commission had calculations showing that these figures could be achieved at relatively modest costs. Yet, in an interview, one EU official later described this as 'what the traffic could bear' – in other words, the targets were also simple focal points (see also Helm 2009: 226). Moreover, the Commission was seemingly somewhat reluctant with respect to the renewable energy target. A Commission interviewee noted that 'this was the only area in which the Council turned out to be more enthusiastic than the Commission.'

Our interviews show that most stakeholders were greatly surprised when heads of state so readily decided adopt the Commission proposal and go for binding targets. Some interviewees give the Commission main credit for the outcome. Others indicate that the central drivers were the high climate ambitions of Chancellor Merkel and Prime Minister Blair, with Merkel in the forefront, since Germany had the EU Presidency at the time. Due to the effective leadership exerted by Merkel, the outcome of this meeting has been called 'the Merkel Miracle' (Claes and Frisvold 2009: 219). The decisions were highly lauded by insiders and observers alike. Commission President Barroso proclaimed that the EU now had 'the most ambitious climate protection strategy anywhere in the world' (*ENDS* 2007b).

From 2006 and onwards, the IPCC presented scientific reports whose main message was that global emissions would need to be reduced by 25–40 per cent by 2020 and by 50–80 per cent by 2050 (see Solomon et al. 2007). The IPCC's Fourth Assessment Report was released in 2007. This contributed to climate mitigation climbing to the very top of the political agenda in many European countries (Jordan and Rayner 2010, Oberthür and Pallemaerts 2010, Wurzel and Connelly 2011b).

Public perceptions of climate change had certainly changed markedly: in 2003, 39 per cent of respondents to a Eurobarometer survey had cited climate change as their main worry: this figure rose to 45 per cent in 2005 and to 57 per cent by 2007 (European Commission 2008a). As regards economic prospects, a 2008 Eurobarometer report noted that 'in spring 2007, the traditional indicators of the standard Eurobarometer reached levels which had not been seen for many years. Europeans were very confident about the outlook for the European economy, their country's economic situation and the employment situation at national level' (European Commission 2008o: 3).

Further policy development in the form of the launching of a specific climate and energy policy package was planned for the autumn of 2007, in order to underpin EU positions and leadership at the global climate meeting in Bali in December. But in October the Commission announced that the proposals would be delayed until mid-January 2008 (*Euractiv* 2007c). So the EU leaders had to travel to Bali with no specific new proposals in their briefcases. In contrast to the increasing global urge for action, international negotiations became mired down in a serious stalemate, due not least to conflicts between the industrialized and the developing countries. Hardly any progress had been achieved at the Conferences of the Parties arranged between 2005 and 2008 (Clémencon 2008, Andresen and Boasson 2012). The Bali meeting resulted in the adoption of the Bali Action Plan to develop a post-2012 global agreement at the Copenhagen meeting in 2009.

Many interviewees report that the stalemate in the global negotiations created a shared feeling of urgency towards developing clear and ambitious EU climate policies. The hope was this would encourage also other parties to strengthen their commitments. The EU had long seen itself as a global climate leader, and the slow pace of international progress had further strengthened this ambition. As one central EU official interviewee reflected, 'Back in 2007 and 2008 there was a war going on. Everyone wanted to finish the package and take it to Copenhagen.' Others have downplayed the specific importance of the climate negotiations and the Copenhagen summit. One EU official claimed, 'Copenhagen was definitely not important. It may have been in some people's minds (...) Only the least realistic people thought that it could result in an agreement (...). Rather, we wanted to show the US that we could do this.' The same interviewee emphasized that 'the price of energy was soaring, we wanted to move our utilities away from coal and towards renewables and nuclear. That was the real motivation.'

In January 2008 the Commission then launched its package of policy proposals. The package consisted of four main legislative proposals for achieving the 20-20-20 targets: 1) a revised EU ETS for 2013–2020 (European Commission 2008g), 2) a decision on effort-sharing between member states in the form of differentiated targets for sectors not covered by the EU ETS (European Commission 2008j), 3) a draft directive on the promotion of renewable energy (European Commission 2008d), and 4) and a draft CCS directive (European Commission 2008n). In addition, new legislation was also underway for CO_2 emissions from cars and for fuel quality.

The package emphasized fairness in burden sharing, and linking climate and energy goals. In general, the proposals were well received by the European Parliament and the Council of Ministers (Skjærseth and Wettestad 2010a). Moreover, a Commission interviewee stressed that 'the package mobilized everybody: Commissionaires, senior officials and member states. It created a dynamic atmosphere'. As regards the concept of a 'climate package', we noted earlier in this chapter that the concept had been introduced already in the early 1990s, but at that time it simply referred to several measures presented at the same time. Others have noted that the 2008 package follows on from experience with monetary union and with the earlier completion of the internal market in the 1992 package (Helm 2009: 228). Our interviewees also drew attention to inspiration from the national climate policy packages that had been developed in Germany and France.

The Commission called for swift deliberations on the package. France was scheduled to assume the EU presidency in the second part of 2008. A well-informed interviewee stated that already in April/May 2008, and before France formally had taken over the presidency, it was decided to get the package adopted at the December 2008 European Council. Our interviewees agreed that this was an important move – and, for many, a surprising one. The rationale was that the EU had to finalize its internal climate policy in due time to maintain its leadership position in the run-up to and at the Copenhagen conference in 2009. As decisions in the European Council are made by unanimity due to its traditional role as the venue for 'history-making decisions' in the EU, this had important implications for the whole decision-making dynamic (Peterson 1995, McCormick 1999: 16). It meant that the legislative proposal did not go through the full co-decision procedure, including a common position in the Council, two rounds ('readings') in the Parliament, and even a possible final round in a Conciliation Committee (see Haigh 2011). Instead, trialogue talks between the Commission, the Parliament, and the Council were intended to sort out the main disagreements in a more rapid, *single* round.[1]

The French Presidency declared the climate and energy package a key priority, and the French worked hard to secure an internal agreement within the EU before the end of the meeting of the global climate regime in Poznan in December 2008. Not only did the French have the necessary capacity, one interviewee also emphasized that the French Presidency was characterized by the attitude that 'if Germany was able to deliver the targets we shall certainly show that we are able to deliver the measures'. However, although the French devoted considerable resources to this task, they lacked some of the technical information needed to ensure the finalization of the package. In what was seen as a highly unusual move, the Director of DG Environment, Peter Mogens Carl, left his post and was recruited into the French EU Presidency (*Financial Times* 2008b).

In the autumn of 2008, the European Parliament's committees for Environment, Public Health and Food Safety and Industry, Research and Energy presented their

---

1   For information on the trialogue procedure, see http://ec.europa.eu/codecision/stepbystep [accessed 20 February 2012].

amendments for the various policies in the package, and parliament rapporteurs engaged in trialogue meetings. The Environment Committee dealt with most aspects of the package, whereas the Energy Committee handled the renewable energy directive. In its activity report, the Environmental Committee summarized the unique process as follows:

> On the basis of deals which were hammered out between the various rapporteurs and the French Presidency in no fewer than 20 trialogue negotiation meetings, the European Parliament was able to approve 6 packages of compromise amendments on 17 December 2008. These first reading agreements constitute in many regards a world record, e.g. in terms of rapidity (finalization of the parliamentary procedure in less than 11 months from the presentation of the core proposals) and complexity of the subject-matters involved; but most importantly these procedures brought the world's most ambitious climate legislation into being. (European Parliament 2009b:15)

As the action-packed European Council meeting took place *prior* to the final plenary meeting in the Parliament, the latter found itself faced with having either to accept or to reject. However, we should also recall that the two committees and particularly the rapporteurs had been heavily involved throughout the trialogue process. The Parliament chose to accept, formally endorsing the deal on 17 December 2008. The Council then formally adopted the package in April 2009. Oberthür and Pallemaerts (2010) note that the speed that characterized the legislative procedure of the package showed the political importance of the issue at the time and the political determination and willingness of EU leaders to achieve an agreement.

Much of the main specific dynamics, conflicts and final compromises struck in negotiating the package are further elaborated in the subsequent chapters in this book. Here let us simply sum up the main outcomes: an improved and extended EU ETS, a new and comprehensive renewable energy directive, and a new CCS directive and accompanying CCS financing mechanisms. In addition, the 'effort-sharing' decision established national reduction targets in the non-ETS sectors, aimed at contributing to the overall reduction target of 20 per cent by 2020. The total reduction target for these sectors was 10 per cent below 2005 levels by 2020, with the effort spread among the member states on the basis of their GDP (Decision 406/2009/EC, see Lacasta et al. 2010).

The EU had then put in place central elements in the preparations for the Copenhagen summit, which was held amidst considerable media focus in December 2009. To cut a long story short: the summit resulted in the rather weak Copenhagen Accord, far from the path-breaking outcome that many had hoped for (see Dimitrov 2010). The meeting was characterized as a huge climate diplomacy failure, not least for the EU. *ENDS Daily*'s report gives a sense of the immediate disappointed response to the outcome:

> The Copenhagen climate summit was a failure on so many levels. The Danish hosts, the UN negotiating system, the big powers, over-ambitious small island states, environmental activists: all must bear a portion of the blame (…) Europe went to Copenhagen as the self-professed champion of an ambitious agreement (…) it has already put into domestic law many of its pledges. Yet in the final analysis all of Europe's grand intentions and environmental commitment counted for little. (*ENDS* 2009j)

The Copenhagen Accord recognized 'the scientific view that the increase in global temperature should be below 2 degrees Celsius'. It did not establish binding commitments, but instead encouraged states to submit emissions reduction targets (UNFCCC 2009).

In the aftermath of Copenhagen we have witnessed deepening financial recession and climate change dropping in importance on the EU agenda. However, it is interesting to note that a 2011 Eurobarometer report found that the European public was more concerned about climate change than it was in 2009 and that climate change remains a greater worry than the economic situation (European Commission 2011e). Reluctance among some countries, particularly Poland, has blocked a move to a 30 per cent reduction target. Much of the attention has shifted to more long-term visions, particularly with the March 2011 roadmap for moving to a competitive low-carbon economy in 2050 (European Commission 2011f). But all this is stuff for a subsequent book.

We have seen that in this phase the EU moved to a new level of climate policymaking. A centrally placed interviewee highlighted that the situation in the late 2000s was truly special: 'We had the perfect conditions for achieving something extraordinary: France and Germany held the presidency at important stages in the decision-making, and the UK was largely positive as well.' The implication is that the decisions of 2007/2008 are not likely to be repeated any time soon: 'Today we would have failed if we had tried to do anything like that. It was partly a miracle and partly a strong presidency.'

Actors, Roles and Positions

From 2005 on, important changes took place in the positions and roles of main actors. Here we will summarize only some striking overall changes and developments, as the more case-specific detailed changes are presented and discussed in subsequent chapters.

We have already noted the increasing attention to and concern about climate change among the public and voters in the EU member states. In particular, many of our interviewees point out that Germany and UK were important in the early phase of the new climate policy drive, from 2005 and until 2007/2008. The latter view is supported by studies of UK and German engagement in EU climate policy, showing that both countries aimed at playing key roles in European as well as global negotiations (Rayner and Jordan 2011, Jänicke 2011). Germany remained a strong supporter of climate policy development, but mounting political challenges for Britain's Labour government hampered its involvement in this context.

There were also important changes in the southern countries. For instance Spain's about-turn in 2004, from being a climate laggard position to taking a far more constructive position, was related to the change of government that year, from Conservative to Socialist (Skjærseth and Wettestad 2008). This change further strengthened the group of governments and countries concerned about climate change.

The new East European member states started to position themselves in EU processes. Due not least to the Russian gas crisis, these countries contributed to the increasing weight given to energy security and the necessity of viewing climate and energy policy in combination. On the other hand, the eastern enlargement of the EU had brought in member states with energy systems dominated by carbon-intensive coal burning and climate change low on their own political agendas (see Skjærseth and Wettestad 2007).[2] Poland in particular stood out as a political heavyweight, with 27 votes in the Council – the same as Spain and only two votes less than Germany and the UK. Poland emerged as a general policy laggard in 2008, and has continued with this basic stance. Several elements in the 'sewing together' of the climate package were intended to accommodate the special needs and interests of the new member states, including the effort-sharing directive and various redistribution elements in the revised ETS.

From 2005 on, the Commission stood forth as an increasingly important policy entrepreneur, as is documented and discussed in greater detail in subsequent chapters. Our interviewees emphasize the excellent collaboration between the Commissioners for Environment and Energy, and between the Directors in the two directorates. But in the process of preparing the various policies in the climate and energy package, there were – not surprisingly – also conflicts both within and between Commission directorates. The case of renewable energy is illustrative, as further elaborated in Chapter 5.

Commission President Barroso did not really come aboard the 'climate train' until 2007. As one interviewee explained it, Barroso, being a shrewd political player, now sensed that the 'political climate for climate' was changing in key member states and that he would need to come up with a response in order to secure a second term for himself. Further, it may well be that the heavy time-pressure and the trialogue process led to greater Commission influence, securing the Commission a more prominent position in this final stage of the process than it would have had in a normal process with two parliamentary readings.

Also the European Parliament was affected by the increasing 'climate hype' and drive in Europe, even though the elections in 2004 brought in an assembly

---

2 The new members in 2004 were the Czech Republic, Cyprus, Estonia, Hungary, Latvia, Lithuania, Luxembourg, Poland, Slovakia and Slovenia. Bulgaria and Romania joined in 2006.

with a more liberal–conservative dominance. Generally, the Parliament supported the new drive in EU climate policy, and came to play particularly interesting roles in the cases of CCS and renewables. Further nuances in roles and positions of the Parliament are discussed in subsequent chapters.

With regard to the role of industries, this new and broad regulatory drive meant an unusually busy time for industry lobbyists and representatives. The broadening regulatory repertoire, not least the CCS issue, and the heightened emphasis on technology development also meant new business opportunities in addition to the multi-faceted regulatory challenges and 'threats'. The more specific industry picture is a central theme of subsequent chapters.

These were busy times also for environmental organizations. They were instrumental in connecting the EU agenda closely to the global climate talks (Wurzel and Connelly 2011c). Various campaigns were set in motion to go from a 20 to a 30 per cent target for the EU, albeit without success. We also note that a range of small and specialized climate NGOs emerged in Brussels, often with a close relationship to industry actors. The environmental groups had generally warmed to 'cornerstone' ETS. Interestingly, in a common position paper on the EU ETS review process in 2007, the Climate Action Network, WWF, Friends of the Earth, and Greenpeace stated that 'the existence of the EU emissions trading scheme (ETS) is a tremendously important achievement for European Climate Change policy' (CAN et al. 2007: 5). They supported the work on increasing the centralization and market streamlining of the system.

Probably the most difficult part of the new drive was the CCS issue. Here, environmental organizations could be found adhering to all the main stances – from the fiercest critic, Greenpeace, to strong proponents among the group of small and specialized actors like Bellona and E3G. This is a clear illustration of the unprecedented broad scope of the new drive, with new challenges for all those involved.

In the following, we proceed to fill in this overview picture with four more deep-going case stories, starting with the cornerstone EU Emissions Trading System.

Chapter 4

The 'Revolutionary' Development of EU Emissions Trading: The Triumph of 'Tortoise' Entrepreneurship?

Introduction

In the 1990s, the EU climate policy debate was dominated by the lengthy tug-of-war between the supporters and the opponents of an EU carbon tax. Eventually the taxation idea failed, but the idea of setting a price on carbon emissions lingered on. Having opposed flexible climate-policy instruments in the run-up to the Kyoto summit in 1997, the EU made a remarkable volte-face after the adoption of the Kyoto Protocol. From 1998 on, it started to develop its own internal emissions trading system (ETS), targeting industry point-sources ('installations') in its member states. In 2003, a Directive was adopted that set out the main rules for a pilot phase of trading (Phase I, from 2005 to 2007), and a subsequent Kyoto Protocol phase (Phase II, from 2008 to 2012) (European Parliament and Council 2003, see Skjærseth and Wettestad 2008, Ellerman, Convery and Perthuis 2010).

Directive 2003/87/EC (European Parliament and Council 2003) established an initially quite decentralized system of in practice 'piecemeal markets', with significant member-state control. Furthermore, as emission quotas ('allowances') were to be distributed free of charge, this market instrument was initially infused with a considerable degree of lobbying and politics. In any case, the ETS emerged as the 'cornerstone' and 'flagship' of EU climate policy, the 'new grand experiment' (cf. Kruger and Pizer 2004). The initial ETS did not function particularly well, with its surplus of allowances and a volatile carbon price – the latter dropping to near zero towards the end of the pilot phase. The flagship seemed lost at sea. However, when the system was revised in 2008, the result was a far more centralized and market-streamlined design for Phase III from 2013–20, with more of a 'Single European Market' character (European Parliament and Council 2009b). Tellingly, the analysts at Carbon Trust have characterized the changes as a 'revolution' in the division of power between the EU and its member states (Carbon Trust 2008: 17).

The two key questions in focus in this chapter are then: Why did the EU choose emissions trading as its flagship in regulating carbon emissions? And why has the ETS ended up as a 'Single European Market' type of system? Earlier studies have noted the key entrepreneurial role of the European Commission (hereafter: the Commission) in the making of this system, holding preferences for a centralized and auctioning-based system but then bowing to industrial and member-state opposition (see Wettestad 2005, Skjærseth and Wettestad 2008). With the revision process, might today's radically reformed ETS be seen as the triumph of 'tortoise' ('slow and steady wins the race') entrepreneurship from a Commission that finally managed to put in place its preferred type of system?

As with all the cases in this book, we will examine how industry, policy interaction and the external environment contributed to shape the Single European Market outcome. Was the Commission's entrepreneurship underpinned by active industry support? Or did the Commission actually manage to have its way, in the face of the economically powerful electricity industry and energy intensive industries and their interests? We will also explore whether interaction with other policy areas has played a role in the policy processes. Has the Commission strategically coupled this issue-area to others in order to enhance its own influence? Third, we will see how the external environment has played into the policy development. Could it be that the Commission actually played less of a creative role than it may seem at first sight, and that today's ETS is simply a result of the EU implementing the Kyoto Protocol? As explained in Chapter 2, we will draw on three approaches – Liberal Intergovernmentalism (LI), New Institutionalism (NI), and Multilevel Governance (MLG) – in specifying the mechanisms through which industry, interaction and the external environment have affected the development of policy.

Main Policy Outcome: Towards a 'Single European Allowance Market'

The ETS is based on the idea that all large point-source emitters of CO_2 must have allowances equivalent to their annual emissions. Around 11,000 'installations' in the power-producing and power-consuming/energy-intensive industries (such as refineries, steel and cement) are targeted (European Commission 2008c). Companies may buy and sell these allowances as deemed necessary. The availability and distribution of allowances are shaped by public regulations, which will influence how the market forces will work and hence what the carbon price will be. Particularly important is the total number of allowances (the cap), which should be low enough to create scarcity in the market. All types of emissions trading systems are based on market thinking, but the degree of market streamlining can vary significantly (see OECD 2006). This applies particularly to allocation methods, where the auctioning of allowances must be seen as a more 'market-aligned' method than free allocations. Also the competence distribution varies. In the case of the EU, both the EU and national governments may in principle have the final say in rule-making and governing of the system.

We begin by briefly summing up the revised ETS steering method, which is to take full effect from 2013. The revisions go a long way towards establishing a Single European Market for carbon emissions. Although the ETS is obviously and fundamentally a market instrument, the initial ETS must still be characterized as an economic-political hybrid (see also Van Asselt 2010: 140). The ETS post-2013 will certainly function differently, with auctioning established as the main allocation method (see section 15 in the preamble to Directive 2009/29/EC, European Parliament and Council 2009b). Some 40 per cent of allowances will be auctioned in 2013, increasing to around 70 per cent by 2020. This means that the distribution of allowances will increasingly be based on market criteria (like willingness and ability to pay), with a corresponding reduction in the influence of technical and political considerations. These changes will reinforce the overall character of the ETS as a transnational market measure.

However, the development is not unambiguous. As a general principle, the power sector will have to purchase all its allowances from 2013 (section 19 in the preamble to the 2009 Directive). Other industries need only to buy at least 20 per cent of their allowances in 2013, increasing to at least 70 per cent by 2020 (with a view to reaching 100 per cent by 2027). Furthermore, certain industrial subsectors particularly exposed to global competition are guaranteed free allowances in the period 2013–2020, to be based on state-of-the-art technology benchmarks (Article 10a in the 2009 Directive).

As regards the distribution of competences, the ETS post-2012 will be a fairly centralized and harmonized system. The 2009 Directive sets a collective target for the ETS as a whole, and national allocations are then to be derived from this single, EU-wide ETS emission cap (see sections 13 and 14 in the preamble to the 2009 Directive). Free allocations will be further harmonized: with inputs from relevant stakeholders, the Commission has produced Community-wide sectoral benchmarks for such allocations. The Commission stands out as a formally important coordinator of the ETS, with for instance a central role in designing rules for monitoring and verification. Furthermore, from 2013 on, allowances will be held only in the central Community registry, not in national registries as well. The Commission is given a coordinating role in the management of and responses to fluctuations in the price of carbons (Article 29a). If a 'satisfactory' new global climate agreement is adopted and the EU raises its overall target from 20 to 30 per cent, the Commission will also have a key role in the subsequent ETS adjustment process (Article 28). Auctioning revenues will be collected by member states, but it is the Commission that is to monitor whether these funds are applied in accordance with the principles of the Directive.

The Commission has a central role in the process of developing detailed regulations for determining which ETS sectors are particularly exposed to global competition and may thus qualify for free allocations (European Parliament and Council 2009b. Directive 2009/29/EC: Article 10a). A clarification process was carried out, mainly in 2009 and 2010, with the final benchmarks adopted in April 2011, by Decision 2011/278/EU (European Commission 2011d).

The upshot of all this is that after 2012, the ETS will no longer be a conglomerate of 27 national emission systems: it will be a harmonized pan-European scheme, with common rules for most aspects – such as how much of the allowances are to be auctioned, which industry activities can still be given free allowances, and how transactions are to be registered.

Add to this the shift to auctioning as the main rule, and we may conclude that the revisions bring the ETS closer to a Single European Market type of instrument. Let us now have a look at the historical development of EU carbon regulation that led to such a result.

The ETS Story: From Subsidiarity and Carbon Market Scepticism to Centralized Market[1]

Pre-1997: Failed Efforts to Develop EU Carbon Regulations

As described in Chapter 3, climate change emerged as a political issue in the EU from the mid-1980s. In the run-up to the 1992 Rio Summit, the EU adopted a commitment to stabilize its GHG emissions at 1990 levels by the year 2000, and then started the search for an effective way of regulating carbon emissions. Initially, the Commission put considerable effort into developing an EU energy/carbon tax. This proved a tough battle, not least because prominent member states rejected the idea: taxation was seen as a member-state prerogative. There was no sign of main OECD competitors establishing carbon taxes, and only a few EU member states adopted national taxes. With consensus looking increasingly unattainable, the idea of a common energy/carbon tax was downplayed towards the mid-1990s.

The emissions trading instrument has a conceptual history that dates back to the 1960s (see Voss 2007, Van Asselt 2010). The main initial practical utilization took place within the context of US air pollution regulation from the 1970s onwards (Stavins 2002, Ellerman, Joskow and Harrison 2003). Emissions trading was discussed in a few EU member states, notably Denmark and the UK, but it remained a peripheral idea. Central European industries such as oil and electricity were cautiously positive. Then the oil majors BP and Shell took initiatives to develop internal emissions trading systems in 1997 and 1998 (Victor and House 2006). Other industries preferred voluntary agreements, most keenly promoted by dominant energy-intensive industries (*ENDS* 1998d). The USA and other parties brought the idea of flexibility mechanisms, including emissions trading, into the international climate discussions, but the EU argued against including such mechanisms in the climate regime (Grubb, Vrolijk and Brack 1999).

In addition to taxation and emissions trading, the EU had another possible way of regulating industrial emissions: setting emissions limits and standards similar to traditional regulations of, for instance, water and air pollution. At this stage, most member states had developed quite elaborate pollution regulations of this kind (see Wurzel 2002). It was becoming necessary to see the regulation of various types of pollutants in a broader, more coordinated perspective. This led to the

---

1   This section has benefited from the ETS process overview carried out in Wettestad 2005, Skjærseth and Wettestad 2008, 2010a and 2010b, and Wettestad 2011. See these publications for greater detail about the various actors and policy sub-processes.

adoption of the Integrated Pollution Prevention and Control (IPPC) Directive in 1996 (see Egelund Olsen 2006, Haigh 2011). Its broad definition covered carbon emissions as well, but climate change was not specifically mentioned or discussed in the IPPC Directive. All facilities covered by the Directive were made subject to permission by national authorities, which meant that member states retained the regulatory upper hand. Given the comprehensive coverage, in principle it would of course have been possible to include carbon regulation more specifically in the IPPC regulatory system. In our interviews, some EU officials mentioned that the Commission had discussed the possibility, but swiftly rejected the idea. As the EU lacked competence in this issue-area, there was no guarantee that the EU would have the clout needed to control national emissions.

We should also note the rather weak renewable policies on energy and energy efficiency at the time – the ALTENER and SAVE policies, respectively (see Chapters 5 and 7). Because these had primarily a symbolic character, there was certainly both room and need for a more ambitious, more tangible EU climate-policy measures. By 1997, then, the EU had not found a viable path towards a common and effective carbon regulation. Most member states had considerable experience in pollution regulation, but this rarely included carbon. Some states had adopted carbon taxes or started to discuss emissions trading, but on the whole there were few pre-existing national traditions here. Moreover, the failed attempt to establish a carbon tax became a symbol of the EU's ineffective policymaking ability in this field.

1998–2004: Trading Turnabout and Shaping of the Initial ETS

A swift and radical shift towards emissions trading occurred in the second phase, after the Kyoto Protocol was signed in 1997. Under the Kyoto Protocol, the EU made a commitment to reduce its emissions by eight per cent in the period 2008 to 2012. Furthermore, the Protocol established three flexible mechanisms – international emissions trading, the CDM and JI – as well as the obligation of showing 'demonstrable progress' by 2005 (Article 3).

From early 1998 onwards, a group of DG Environment officials, most of them economists by training, began campaigning for the development of an EU emissions trading system (ETS). Unlike the case of the carbon tax, an ETS could be adopted by qualified majority voting. These DG Environment officials certainly played an important entrepreneurial role: one of the authors of this book has earlier referred to the group as the 'BEST group' – 'Bureaucrats for Emissions Trading' (Skjærseth and Wettestad 2008). The BEST group managed quite quickly to convince Environment Commissioner Ritt Bjerregaard of the potential of the emissions trading instrument. Emissions trading was pitched as 'something for everybody': in addition to cost-effectiveness and the instrument's relative popularity among industry, emphasis was placed on how cap-setting and good monitoring provided central control and guarantees of environmentally-friendly outcomes. This versatility helped in overcoming reluctance elsewhere in the Commission, and a relatively high degree of internal consensus was achieved.

In the further development of the ETS idea, the BEST group commissioned and utilized reports on design dimensions and experiences from European and US consultants. The March 2000 Green Paper on EU emissions trading showed that the Commission favoured a design with similarities to the quite centralized air pollutants trading system of the USA (European Commission 2000b). The paper signalled a basic preference for a centralized approach in determining the total amount of allowances (the cap) and the allocation of these allowances, warning that a decentralized approach might lead to national differences and an uneven economic playing field. Furthermore, it seemed reasonable to keep the scope rather narrow. The main focus was on electricity and heat production, with five other significant industrial sectors also singled out.[2] Auctioning was the preferred method of allocating allowances, based on standard economic textbook wisdom.

Industry, as well as EU member states, responded to the Green Paper. Three central industries were electricity, oil (here: mainly refineries), and the energy-intensive industries. The two latter industries participated in global markets, but the electricity industry had primarily a national character at the time, although some of the largest utilities had started to Europeanize their activities (see Chapter 5). Moreover, liberalization of European electricity production was in its infancy, so the electricity industry was not as clearly aligned to a market economic logic as the two other industries. Oil and electricity were certainly supportive of market instruments and the application of the Kyoto mechanisms, but less clear as to whether market measures should be developed at the national or the EU level. As noted, the oil corporations BP and Shell put in place internal emissions trading, and the euro-federation of the electricity industry, EURELECTRIC, carried out emissions trading modelling exercises in 1999 and 2000 – GETS I and II (EURELECTRIC 1999, 2000). As stated by EURELECTRIC, the ETS would 'help European industry to acquire valuable experience in anticipation of an international trading system starting in 2008' (EURELECTRIC 2000).

However, the energy-intensive industries maintained a fairly cautious stance. For instance, the steel industry's euro-federation EUROFER put forward an ETS position paper in 2000. Here, EUROFER noted that although 'emissions trading *could* provide a flexible means for companies to achieve their targets ... the starting point [should] be the global competitiveness of the steel industry [and] a cap on the absolute level of CO2 emissions from the sector or a steel company could distort competition' (EUROFER 2000, emphasis added). Particularly the chemicals industry was reluctant. With regard to structure, these industries were less coherent as a group than the case of utilities. For instance, chemicals and industry metals relate to different global markets; they are represented by different business federations in Brussels; and they are not as dominated by a handful of large corporations as in the case of utilities. Thus, neither the dominance of a few really large corporations nor strong business associations could contribute to create internal agreement. As

---

2   The five were iron and steel; refining; chemicals; glass, pottery and building materials; and paper and printing. European Commission 2000b: 14.

to institutional logic, the energy-intensive industries are characterized by market logics, but with a global outlook – not predominantly national or European, as in the case of utilities. Thus, their growth strategy involves securing national and EU policies unlikely to challenge their global competitive situation.

Furthermore, the electricity industry and the energy-intensive industries were central actors within what can be seen as two separate European organizational fields. The electricity industry was dominated by the seven major corporations and was tied to many national governments, although with rather loose links (see Chapter 5). As regards EU institutions, the electricity utilities had good contacts with DG Environment. The organizational field of energy-intensive industries was characterized by greater diversity and less unity among corporations. With regard to links to EU institutions, these industries were the first ones to be targeted by the EU (as with the original coal and steel union), so they had close ties to DG Enterprise. Although the EU internal market provides the key regulatory framework for these industries, they have also been able to draw on historical ties to national governments, as in the case of voluntary agreements in Germany (Wurzel 2008).

When it comes to the positions of the member states, we can identify three main groups of countries: a small group of emissions-trading supporters, with the UK in the forefront; emissions-trading sceptics, with Germany as the main representative; and the undecided – the largest group. In the UK, a carbon policy shift took place with the election of the Blair Labour government in 1997. Greater attention was paid to market-based instruments, leading to the adoption of a national emissions trading scheme of a voluntary character in 2002 (Jordan et al. 2003). The UK established itself as an emissions-trading frontrunner and supporter, but it favoured a decentralized EU ETS in order to ensure compatibility with its own domestic system.

As to the industrial basis, none of the European major utilities were British, although several had a strong presence in the UK market (HM Government 2007). The overall positive position of the UK was certainly backed up by dominant sentiments within British industry. For instance, as early as in 1998 several industrial bodies published reports that were positive to the introduction of emissions trading; and a specific trading group was formed in 1999 (Jordan et al. 2003: 189, Wurzel 2008: 11, Skjærseth and Wettestad 2008: 88). The British oil industry, with BP and Shell at the fore, supported the development of a UK emissions trading scheme. Also Sweden welcomed the EU ETS with some enthusiasm – and Sweden did not have a national scheme to protect (Oberthür and Tänzler 2007). It was also home base to Vattenfall, a major European utility.

Germany stood out as the prime sceptic to emissions trading. Given the size of its economy and the magnitude of the related emissions, the country was destined to be one of the truly key players in the ETS. In the late 1990s, Germany's climate policy mixed traditional regulation with voluntary agreements and eco-taxes (Wurzel 2008: 13, Watanebe 2011). Germany is home to several major European companies, among them the steel giant ThyssenKrupp, the chemical giant BASF and the big electricity utilities E.ON and RWE. Voluntary agreements sat well with the majority of German industries, which were generally opposed to emissions trading. Particularly the energy-intensive industries stood forward as emissions-trading sceptics. The German government called for a decentralized EU ETS in which several installations would have the possibility to adopt aggregate, collective targets. Central actors, including the key Economics Ministry and Prime Minister Gerhard Schröder himself, were ETS sceptics (Wurzel 2008: 14). This was probably founded in a basic responsiveness to industrial concerns, mixed with regional politics.[3]

Spain can be regarded as a representative of the third and largest group of countries – the undecided – that took a bystander position. At that time, Spain had a conservative government, and climate-policy issues – including emissions trading – were accorded scant attention or priority (Wettestad and Sæverud 2005, Del Rio 2007, Costa 2011). This resonated well with main sentiments within Spanish industry, which also considered their country's target under EU's burden-sharing agreement to be unfair. Spain was home base to two big European utilities, Endesa and Iberdrola. France and Italy held positions similar to that of Spain, but also many smaller EU member states may be counted to this 'undecided' group. At that time, Poland, along with other Central and Eastern European Countries (CEECs), had only observer status in the process of establishing the initial ETS. Poland had no major utilities, its market dominated by smaller Polish utilities (PAI 2006).

Let us again take up the threads of our chronology. The BEST group referred to the flexible mechanisms that were adopted in the Kyoto Protocol to legitimate the development of an EU system, for instance by arguing that an ETS from 2005 would 'provide invaluable practical experience of trading' (European Commission 1998b: 20). The US withdrawal from the Kyoto Protocol in March 2001 provided an extra opportunity for playing the international carbon leadership card. Now Commission entrepreneurs argued that, in order to bring the international process forward, the EU would have to develop significant internal policies. The global situation enabled these entrepreneurs to frame the ETS as a core instrument for both bolstering EU global leadership and saving the Kyoto Protocol as well (see Cass 2005).

The BEST group led a working group under the first European Climate Change Programme (ECCP I), and used this multi-stakeholder forum as a formal platform for establishing a network positive to trading (Skjærseth and Wettestad 2008: 82–83). According to one EU official we interviewed, 'the process in the early 2000s can be described as an intellectual achievement: we had to explain the idea and the theory behind it all.' The BEST group used a range of subtle steering instruments, such as agenda setting, participation, and formulation of proceedings.

---

3   Wurzel (2008: 15) mentions the importance of regional/local politics as well: 'The chemical industry and IGBCE (the coal, mining, chemicals, and energy union) found it easy to lobby Chancellor Schroder and Economics Minister Clement because the chemical and coal industries provide many jobs, particularly in the North Rhine Westphalia.'

In the Programme, the ETS was discussed in Working Group 1 (on 'flexible mechanisms'), whereas the two others were discussed in other groups. The ETS was treated specially, as one track of what the Commission referred to as a 'twin-track approach' (the ECCP being the other track).

On this backdrop, in 2001 the Commission then put forward a proposal for a new directive that differed from the Green Paper. First and foremost, it was proposed to establish a system with a fundamentally decentralized approach to setting emission caps. These caps and the more specific distribution of allowances in each member state were to be specified in national allocation plans. The Commission was relegated to a sidelined, watchdog role, primarily the extent to which member states adhered to the common guidelines set by the Directive. Furthermore, a system based on free allowances was proposed. This can be seen as a response to significant scepticism, among member states as well as industries, to the prospect of allowance auctioning.

In other respects, however, the 2001 directive proposal was in line with the Green Paper. The system would start with an initial pilot phase from 2005 to 2007, followed by the second, Kyoto commitment phase from 2008 to 2012. A narrow and limited initial system was proposed, targeting only CO_2 and 'energy activities', which represented by far the main part of regulatory action, although also steel, cement, glass and pulp and paper industries were covered. Links between the ETS and the Kyoto Protocol's flexibility mechanisms – the Clean Development Mechanism (CDM) and Joint Implementation (JI) – were seen as desirable. Due to uncertainties about the final rules of these mechanisms, however, a separate subsequent linking proposal was suggested.

The relationship to other policies did not figure as a central element. The direction of (potential) interactions was on the whole quite unspecified. In the 'Explanatory Memorandum' to the directive proposal, it was noted that the ETS should be 'compatible with the liberalisation of energy markets' (European Commission 2001c: 6).[4] The relationship to the IPPC and Large Combustion Plant directives was mentioned in similarly general terms. Moreover, the policy discussions on renewable energy policy and energy policy for buildings underway at about the same time were not explicitly related to the ETS. Most notably, the Explanatory Memorandum touched on links with 'renewable certificates', and stated that member states 'should take account of renewable energy targets when deciding on the quantities of allowances to be allocated' (European Commission 2001c: 16).

Since important concerns and sentiments among the member states had been integrated into the directive proposal, no radical changes were proposed in the ensuing processes. What then was the role of the European Parliament? It pushed for greater centralization as well as market streamlining, by proposing a

---

4   Also interactions with energy taxes (point 7), environmental agreements (point 8) and the IPPC Directive (point 9) were briefly discussed, generally emphasizing the need for synergistic relationships.

common EU ceiling (cap) on the total amount of allowances, and stood forward as a proponent of significant auctioning. However, it was not able to persuade the member states to change their stance on these points.

Key elements of the proposal – decentralization, free allocation as a ground rule, and the narrow scope – were not altered in the decision-making process, thereby becoming central features of the final 2003 ETS Directive, adopted in July 2003 (European Parliament and Council 2003). The character of this proposal contributed to securing a Council majority that backed the establishment of an ETS. Although free allocation was the ground rule, a limited amount of auctioning was possible: maximum five per cent in the pilot phase and ten per cent in the 2008–2012 phase. The chemicals, aluminium and transport sectors were mentioned as specific candidates for future inclusion in the system. This and other issues were to be further discussed in a review of the system, to be started in 2006. A process to specify the links to the global flexibility mechanisms immediately followed the 2003 Directive, and was finalized in April 2004 (European Parliament and Council 2004, Flåm 2009). A main outcome was the possibility to use CDM credits during the pilot phase, with also JI credits to be used from 2008.

Although the initial ETS was essentially a market instrument, it can still be characterized as something of an economic-political hybrid. The 2003 Directive entailed the establishment of an initially quite decentralized system, with significant member-state control. Furthermore, the distribution of allowances without cost meant that this fundamentally market instrument was initially infused with a considerable degree of lobbying and politics. Hence, in our terms, the initial ETS is best characterized as a 'piecemeal markets' policy.

2005–2010: Initial Malfunctioning and Significant Revision

The piecemeal markets ETS was then launched as planned in January 2005. Member states had allocated generous amounts of allowances to industry, so expectations were low as regards any significant scarcity in the market and corresponding high allowance prices (see Ecofys 2004, Grubb, Azar and Persson 2005, Ellerman and Buchner 2007). Contrary to these expectations, allowance prices climbed to surprisingly high levels from mid-2005 on, peaking at around 30 euros in July 2005. This was probably due to the fact that many (small) companies were slow to enter this new market, creating an artificial impression of scarcity. The high allowance price was accompanied by rising electricity prices. Some claimed that the latter were caused by the former, and energy-intensive industries complained that the power companies were reaping windfall profits: first they received emissions trading allowances for free, and then they went on to make profits from increases in electricity prices related to precisely the introduction of emissions trading (Wettestad 2009a). This discussion continued throughout the pilot phase.

However, ETS dynamics changed dramatically in the spring of 2006, when it became known that four per cent *more* allowances had been handed out than what was needed to cover actual emissions in 2005 (Ellerman and Buchner 2007). This indicated that the whole pilot phase was 'over-allocated'. The allowance price was immediately halved – and continued to drop to almost zero in 2007. The allocation of allowances for the second phase of the ETS (2008–12) was carried out mainly in 2006. It produced national allocation plans that were less generous, due largely to the tougher watchdog line adopted by the Commission. For instance, significant cuts were made in the allocation plans of Germany and not least Poland (Skjærseth and Wettestad 2008, Wettestad 2009b). Furthermore, a new element in this second phase was that allowances could more easily be banked to subsequent phases, potentially stabilizing allowance prices.

The Commission started work on the ETS review in the autumn of 2005, initiating, among other things, a web survey on the ETS for government officials and non-governmental stakeholders. A special multi-stakeholder working group under the second European Climate Change Programme (ECCP II) was to prepare recommendations for a revised ETS. As a linked development, the process of extending the scope of the ETS to aviation had started, and a first Communication on the subject was issued (European Commission 2005b). Due to its limited importance for the steering method and competence distribution dimensions in focus here, we will not go further into this particular process. A strong impetus for ETS reform came with the Commission's November 2006 Communication on 'Building a global carbon market' (European Commission 2006a). The malfunctioning of the initial system had made it easier for the Commission to campaign for the development of a more market-streamlined and more centrally steered system.

Climate change was now rising on the political agenda of the EU member states and institutions. In March 2007, the EU leaders adopted the '20-20-20' targets, with a 20 per cent reduction of greenhouse gas (GHG) emissions as the most relevant for the ETS (European Council 2007). Against the backdrop of new targets and a new dynamic in EU climate policy, four stakeholder meetings were held during 2007 in the working group on the ETS reform under the ECCP II. As to the scope of the ETS post-2012, a cautious broadening to include more sectors and more gases was seen as a natural development.

With regard to the key issues of competence distribution and harmonization, sweeping changes had indeed taken place in the positions of member states. At a specific working group meeting on this in May 2007, a strong call for greater harmonization was evident, although not necessarily for a centralized EU cap (European Commission 2007d). Core states like the UK were highly dissatisfied with the differing implementation of the ETS and the related uneven economic playing field. Participants from the new EU member states emphasized the need for some continued flexibility, in order to accommodate differences in economic development and the possible impacts on economic growth. As to the method of allocation, several member states spoke out in favour of auctioning (European Commission 2007d: 19). But within industry there was also considerable opposition and hesitancy towards auctioning, particularly among the energy-intensive industries. A key EU official interviewed noted that 'still no one really had any alternative. The industry did not want regulation and they did not want a tax.'

In January 2008, the carbon price had increased again, and allowances were sold for around 22 euro. As noted in Chapter 3, economic prospects were still good at this point in time. After intense lobbying in late 2007 and early 2008, not least from the energy-intensive industries, the Commission put forward its ETS reform proposal on 23 January 2008 (European Commission 2008g). As noted earlier, changes that have been characterized as 'revolutionary' were proposed. First, instead of the previous conglomerate of national allocation plans, the Commission proposed to set a collective target for the ETS as a whole and to derive national allocations from this central cap. The level of the EU-wide cap was to be calculated on the basis of the target for 20 per cent reductions in GHG emissions from 1990 by the year 2020, adopted back in 2007. This pointed towards the need for a 14 per cent reduction by 2020 from the level of emissions in 2005. The linear reduction consistent with this target amounted to 1.74 per cent per year, which would yield a specific reduction of ETS emissions of 21 per cent below 2005 emissions by 2020, and 10 per cent for the non-ETS sectors.

The second important change proposed was that auctioning should be the main principle for all allocation. However, a key differentiation was proposed between installations 'engaged in electricity production' and 'industrial installations'. With regard to the former, full auctioning would be the rule from 2013 onwards, taking into account the ability of power producers to pass on the increased costs of CO_2 emissions (European Commission 2008g: 15). For the energy-intensive industries, a transitional system was envisaged, where the amount of free allocations would be gradually reduced from 80 per cent in 2013 to zero by 2020. Certain industries and subsectors that were deemed particularly vulnerable to international competition and that met relevant criteria could be allocated allowances free of charge, up to 100 per cent. Such subsectors were to be identified by mid-2010. No new CDM or JI credits were to enter the system (only those banked from the 2008–2012 phase) unless a 'satisfactory' new global climate agreement could be agreed and a move to the more ambitious 30 per cent EU goal took place.

The Commission entrepreneurs could now more openly and with greater weight put forward their preferences for a centralized, auctioning-based system – views that they had held already around 2000. The atmosphere of urgency created in the run-up to the 2009 Copenhagen climate summit enabled the Commission to press especially hard for a development in this direction. It was clear that the ETS reform needed to be completed before the seminal Copenhagen meeting, and time was even tighter than in the process that had led up to the 2003 Directive.

By summer 2008, the preliminary discussions had outlined some main positions. This is then a good point for us to pause and sum up the important changes that had taken place in the positions of member states and industry. On the whole, the member states had become considerably more positive to a more harmonized ETS. The UK continued to stand out as a frontrunner, and also Sweden was generally positive. Spain had become far more constructive and positive. Germany had also become more positive, but was still something of a champion for favourable treatment (read: free allowances to) of energy-intensive industries. Poland, now fully involved in the negotiations, stood as the main obstacle. Poland as well as other CEECs feared that auctioning could threaten the economic viability of the many coal-fired power stations in these countries. In other words, an East–West dimension had become manifest in the ETS reform process (see *EU Energy* 2008, *Point Carbon* 2008a, *Reuters Planetark* 2008a).

Several shifts had also taken place with regard to industry positions. This time round, the oil industry played a less visible role in the ETS deliberations. The electricity industry had become more Europeanized and more deeply characterized by market thinking. By 2008, some two-thirds of European electricity sales were controlled by seven majors: EDF, E.ON, ENEL, GDF Suez 2009, Iberdrola, RWE and Vattenfall.[5] Most of the companies had grown considerably due to country-internal mergers. All except Sweden's state-owned Vattenfall were among the 50 highest valued publicly traded corporations in Europe (*Financial Times* 2008a).

Initially most of the major utilities had been governmental agencies, but now the majority had become publicly traded companies keen to increase their market value – and hence firm believers in market measures (Glachant 2003, Rademaekers, Slingenberg and Morsy 2008). In line with this, the electricity utilities had become explicitly supportive of centralized steering of the ETS (EURELECTRIC 2007: 14). They had also become increasingly positive to more auctioning and further market streamlining. But on this issue, differing views within EURELECTRIC were acknowledged (EURELECTRIC 2007: 16). The 'nuclear and renewables' faction was more positive to the shift to auctioning than was the 'fossil fuel' faction. Poland's coal-based electricity producers were particularly vocal opponents to more auctioning (Skodvin, Gullberg and Aakre 2010: 863–64).

The energy-intensive industries were still fundamentally sceptical to auctioning. But they had now become more open to cap-setting at the EU level and to more centralized steering of the ETS (Key Stakeholder Alliance for ETS Review 2007). As to links to the EU bodies, these industries were held to have developed closer links to the industry committee in the Parliament than to the Environment Committee (*EU Observer* 2009).

As an important decision-making clarification, the French EU Presidency had decided that the ETS reform would be decided by consensus at a European Council meeting scheduled for December 2008, together with the other elements of the package. Standard procedure would have been co-decision, with the possibility of majority voting in the Council of Ministers. On the one hand, the broader package context and high-level political involvement meant in principle good possibilities for political deals and horse-trading. On the other hand, the consensus requirement would provide individual laggards with additional veto powers. We return to this in the subsequent analysis.

---

5   Authors' own calculations, based on company annual reports for 2009.

The next central step in the process was ETS reform debate in the Parliament's Environment Committee in early October (European Parliament 2008a). The Committee gave support to the basic reformed ETS architecture proposed by the Commission as regards centralization, the move towards more auctioning, and the rather restrictive line on the use of external credits. However, the Committee also called for *all* auctioning revenues to be earmarked for climate-related purposes, and for up to 500 million allowances from the new entrants' reserve to be set aside as a funding mechanism for CCS projects.[6] The ETS reform process was now entering its final and decisive decision-making month – against the backdrop of a worsening financial crisis. Most of this phase consisted of trialogue meetings and negotiations involving representatives from the Commission, Parliament and Council.

Let us briefly recall the main elements of the European Council agreement in December 2008 and the subsequently adopted revised Directive of 2009. First, with regard to the cap, it was adopted along the lines proposed by the Commission in January 2008: a single, EU-wide cap from 2013 on, and allowances to be allocated on the basis of fully harmonized rules. Second, significant elements of the outcome on allocation were in line with the Commission proposal: auctioning was introduced as the main allocation method, combined with an initial differentiation between power producers and energy-intensive industries.

But new elements were added to the outcome, as compared to the Commission's original proposal. For instance, 300 million allowances from the new entrants' reserve were set aside to support up to twelve CCS demonstration projects and projects demonstrating renewable energy technologies. In addition, the member states were *recommended* to use 50 per cent of the revenues on measures to fight and adapt to climate change mainly within the EU, but also in developing countries (Article 10(3)c).

What then about policy interaction in the ETS revision? We see a notable formal link to renewables, which were negotiated as part of the same package. However, the two processes were still clearly separated, with the ETS review carried out under the auspices of the ECCP II 'working group on ETS reform'. Furthermore, DG Environment was in charge of the ETS review, whereas DG Energy chaired renewables. Also separate impact assessments were carried out, in addition to a combined one (see European Commission 2008 l, m). We do not find any examples of actors who argued that the ETS should learn from or be influenced by renewables policy design: it was more the other way around. The energy policy for buildings was revised mainly in the aftermath of the climate package, and there were no evident links between this and the ETS development. Concerning CCS, this policy area was hardly linked to ETS in the initial policy preparation phase. CCS was not mentioned at all in the 2003 Directive. The link

---

6   The new entrants' reserve is a pool of allowances for companies starting up production in the allocation period.

between these two policies developed substantially in the period 2006–2008, as will be further discussed in Chapter 6.

Further clarification of the issues of carbon leakage and industrial benchmarks took place from 2009 to the spring of 2011. A list of vulnerable sectors was published in the autumn of 2009, with 164 sectors specified (European Commission 2009a). In June 2010, the Commission then announced that it was working on 53 product benchmarks for 20 sectors *(ENDS* 2010). Draft benchmarks were put forward in December 2010, with the final benchmarks adopted in April 2011 (Decision 2011/278/EU, European Commission 2011d). The steel industry was particularly dissatisfied with the benchmarks that were adopted, and in July 2011 initiated action at the European Court of Justice to get the Commission decision annulled (EUROFER 2011). At this point in time, the financial crisis and reduced need for allowances had put the discussion about further ETS revision and particularly a tighter cap on the agenda. The revolution had seemingly not been radical enough.

Let us now examine how the industry, interaction and EU-external mechanisms laid out in Chapter 2 have shaped the ETS outcome.

Why has the EU Moved Significantly towards a 'Single European Allowance Market'?

Industry: Significant, but not as Storm Troops for the Revolution

We begin by asking how carbon regulation as a policy area has been shaped by issue-internal dynamics within the EU, with a specific view to the role played by industry. In the 1990s, the idea of regulating carbon through a carbon tax was shelved, due not least to lack of industry support. Above, we noted two prime groups of industrial actors in this case: electricity utilities, and various energy-intensive industries. In pre-1997 discussions on carbon regulation it is hard to discern any sharp divisions between the positions of these two groups. Later, they came to perceive their economic interests quite differently and put forward different main positions on EU carbon regulation. In general, the electricity producers have been quite positive to emissions trading, over the years becoming increasingly positive to a 'Single European Market' design as well. By contrast, the energy-intensive industries have been far more sceptical towards this instrument, gradually becoming more vocal in their calls for another type of system – a baseline-and-credit system, centred on sector benchmarks.

Moreover, the electricity producers were active in both policy development processes, whereas the energy-intensive industries became more involved after 2004, especially in connection with the windfall profits that the utilities were said to be reaping. They perceived that the initial ETS was entailing higher costs for them, also indirectly by increasing electricity prices linked to the ETS. Few utilities, whether coal-based or not, experienced increased costs from the initial 'piecemeal market' ETS, but the utilities were dissatisfied with the uneven framework conditions created by this system. The 2009 Directive contained elements that arguably re-dressed the balance between the industries a good deal. The 'close to Single European Market' outcome is closer to the main positions of the electricity utilities than the energy-intensive industries, but the latter managed to secure significant portions of continued free allowances up to 2020 and to introduce benchmarking. How can our different perspectives and mechanisms shed light on this?

As regards the *instrumental governments* mechanism, it seems clear that it was *not* the member states responding to the demands of industry that placed the ETS on the EU agenda: the initiative came from the Commission. However, neither utilities nor energy-intensive industry rejected the idea outright. It is also clear that the 'piecemeal markets' 2003 Directive was significantly flavoured by shared positions among industries. Particularly the energy-intensive industries, but also the utilities, preferred a decentralized structure with allowances to be handed out for free, as against the Commission's preferences for a centralized, market-streamlined system.

Did the member states in focus here align with their national electricity industries or energy-intensive-industries at the time? Germany was home base for giant companies in both categories, and the empirical backdrop indicates that both industries were able to influence the German positions, but the sceptical position of the government seems to have been tinged just as much by widespread ideological scepticism towards emissions trading as direct industry influence. Spain harboured two large utilities, Iberdrola and Endesa, but also some energy-intensive industries. However, the government's positions seem to have stemmed primarily from non-engagement in climate-policy development in general. The Spanish government simply did not pay much attention to EU carbon regulation or how it should be designed.

Sweden was home to big (and state-owned) utility Vattenfall and hosted also some notable energy-intensive industries, for instance in the steel and pulp/paper sectors. It is likely that Vattenfall's dominant position contributed to the government's positive approach to emissions trading. The UK is more of a special case here, with oil companies as the prime industrial giants, and the lack of a strong presence from either of the two other industries. It seems quite likely that the initial development of a UK ETS resulted partly from oil industry influence, making Whitehall fairly positive to this climate measure. However, in the decision-making process leading up to the 2003 Directive, the British government was concerned mainly about the match between its own domestic ETS and the EU system, and did not act as an instrument for the oil industry. On the whole, then, we may conclude that, in the initial ETS process, the positions of many EU member states were influenced by their national industries, but it does not make sense to see them as acting solely as instruments for the most dominant national industries.

To what extent did the governments act as instruments for the industries in the process leading up to the 2009 Directive? To some extent, highlighting central industrial dynamics and pressures in key member states can provide explanatory insights here. Germany is probably the most prominent case. Its energy-intensive industries stepped up their engagement and apparently succeeded in shaping the German 'campaign' for continued free allowances to energy-intensives in the ETS post-2012. However, since we do not see a clear shift in the economic weight of the two industries, a Liberal Intergovernmentalist (LI) perspective cannot provide a full explanation for Germany's greater alignment with the energy-intensive industries. The UK, Sweden and Spain all increased their support to market streamlining and centralization; and, although this was in line with the shift in the positions of the utilities, it does not seem that their shift in attitude was due to the governments merely acting as instruments for their own national industries.

Moreover, we can note at least two central differences between the deliberations in 2002/2003 and in 2008. First, the enlargement process had added 12 new member states to the EU. This development served more to impede than support the 'revolutionary transition' towards a Single European Market system. A central concern for these countries, led by Poland, was to water down the suggested auctioning of all utilities allowances – and in this they had some success. The Polish government, more clearly than that of any of the other member states, did act as an instrument for nationally dominant industries – its coal-based energy industry – so that it fiercely supported exemptions to the Single European Market design. Second, as the revised ETS Directive was adopted at the European Council level, not the Environment Council as in 2003, those member states who were opposed to auctioning acquired blocking veto powers. This change in formal procedure was promoted by countries such as Poland and strengthened their ability to promote the interests of coal-based utilities and energy-intensive industries.

We can conclude that, as projected by the LI approach, the economic interests of industries have indeed been central in the development of EU emissions trading. There are bits and pieces of evidence to indicate that an instrumental government mechanism was in operation in the process leading to the original 2003 ETS Directive. Yet, this perspective cannot explain why the energy-intensive industries were quite passive at this stage. Furthermore, the system was certainly not initiated as a response to strong demands from industry. In addition, the 2009 'revolution' was not really initiated by industry, although it was supported by most utilities. Instead, when the energy-intensive industries succeeded in influencing governments towards the end of that decade, this served more to hamper than to underpin the development of a fairly centralized, market-streamlined system. Hence, an instrumental government dynamic contributed more to create *exceptions* to a fully-fledged Single European Market outcome than in creating the 'revolution' itself.

What then of the role of *institutional feedback* mechanisms? After 1997, the main industries in this case came to relate to two separate organizational fields, with differing views on the ETS and its design. The liberalized electricity utilities were central in the field of European electricity production. The UK and Swedish governments, which had liberalized their energy markets and explored the development of emissions trading at the national level, can be seen as affiliated with that field. However, privatization and liberalization of the utilities made the ties between the utilities and the national governments weaker during the first decade of the new millennium, and few governments were deeply involved. The field can be seen as encompassing parts of the Commission, such as DG Competition and the ETS architects in DG Environment (later DG Climate), and was increasingly embedded in a market logic. First, liberalization of electricity regulation turned the large utilities into increasingly shrewd market players. Second, through modelling exercises, the utilities became accustomed to this new mode of carbon regulation early on. Moreover, they realized they could 'pass through' the costs of emissions trading by means of higher electricity prices – and, since allowances were handed out for free, this would enable them to reap considerable windfall profits.

We see some exceptions to this main trend, in the form of an emerging internal division between groups of utilities with differing energy portfolios. For instance, Poland's coal-reliant utilities were hardly liberalized and stood forth as firmly opposed to more auctioning. But the overwhelming market dominance of a handful of large utilities, and a market-positive European business association, gave the market logic an increasingly strong position. Moreover, support to the ETS instrument was growing within the organizational field of European electricity production. Note also that by 2006/2007 the member-state governments had invested considerable energy, administrative resources and political prestige into developing this system, so it was only natural that these national governments should show greater support to the market logic that underpinned the ETS.

As for the energy-intensive industries, these were initially far less unified than the electricity industry. They were a group of varying industries with different markets, rooted in different home countries and represented by a whole array of different European business associations. Initially, they did not develop a clear and coherent common position. However, most of the sub-industry groups had become firmly embedded in market logics already by the early 1990s; and these actors were central in EU internal market and competition regulation policies. True, they expressed criticism of the increased costs that the ETS could entail, but did not campaign strongly against the idea in the period 1997 to 2004.

We see a radical shift after 2005, when the various industries 'woke up' and began to improve their collaboration. At this stage they were also able to mobilize the larger European organizational field of energy-intensive industries. After all, the industry actors had strong ties to central governments (Germany in particular), but also to strong Commission DGs (such as DG Enterprise). Due to increased internal field-level collaboration among industries, EU officials and national governments, the industries succeeded in attracting considerable attention to both an 'EU-internal anomaly' (windfall profits), and increasingly an 'external anomaly' (carbon leakage) (Wettestad 2011). The industries now experienced that the utilities passed the extra costs related to the ETS over to them, and because they participated largely in global markets (not regional or European markets, as with the utilities) they could not pass higher product prices on to their customers in a similar way as the utilities. Hence, the initial ETS led to greater industrial hostility towards auctioning and more market streamlining within this field – basically the converse of the effect in the European field of electricity production.

We see two contrasting institutional feedback effects at work here. First, developments within the organizational field of electricity production strengthened the Commission's case for more of a purely market design and greater centralization of control. Somewhat paradoxically, the initial malfunctioning of the scheme led both utilities and member-state governments to call for improvements of the existing system – with greater market streamlining and centralization – rather than switching to alternative regulatory routes. The Commission had been given a sufficiently prominent position in the first ETS directive to ensure a strong hand in the revision process. For instance, the 2003 Directive had formally stated that it was a review report to be produced by the Commission in 2006 that would serve as an important reference point in the further development of the system. This helped the Commission to facilitate the revision process in a way that underpinned the growing support towards an ETS with a Single European Market character. And the debate about windfall profits made it harder for the utilities to maintain their initial reluctance towards auctioning.

Second, we see somewhat opposite feedback effects within the organizational field around energy-intensive industries. Experience with the first ETS phases led this field to unify in calling for exemptions to market streamlining of the instrument. Interestingly, even though some central industrial actors saw it in their interest not to be included in the scheme at all (chemicals, for instance), these field actors did not challenge the existence of the ETS as such. This must be understood as a consequence of the prestige of the ETS as the cornerstone of the EU climate policy: the question of abolishing the system was simply not on the table in deliberations now. Moreover, the energy-intensive industry actors were not able to launch a credible alternative to emissions trading. The proposed baseline-and-credit design did not attract widespread interest and support, whether among the utilities or EU officials. On the whole, the feedback effects from the initial ETS within the organizational field of electricity production created greater pressures towards a Single European Market design than the countervailing feedback effects within the field of energy-intensive industries.

In summary, then, how important was institutional feedback in the development of the ETS? First, the actual emergence of the ETS cannot be explained by such a mechanism. Yet, it seems as if the relative success of electricity utilities over energy-intensive industries with regard to the ETS design in 2009 came about partly as the result of a sort of historical feedback within the respective organizational fields. As New Institutional lenses help us to see, decisions made early after the turn of the millennium changed how the industries pursued their interests and how policy was made (Fligstein and Stone Sweet 2002: 1219). Furthermore, with social actors making commitments based on earlier decisions, any exit from existing arrangements became unlikely (Pierson 1996: 146).

The system established in 2003 had an almost trail-blazing quality. Because the organizational field of energy-intensive industries had failed to mobilize strongly in the pre-2003 policy processes, these actors lost out when it came to shaping the fundamental character of EU carbon regulation. By 2008, all they could hope for was exemptions: they had no way of challenging the powerful institutional feedback forces pointing towards a Single European Market design. That said, we must also take entrepreneurial factors into account in order to understand how the energy-intensive industries could make carbon leakage and the related need for continued free allowances such an important element of the revision process.

Turning then to the *network entrepreneurship* mechanism, we can readily conclude that Commission officials, not industry actors, were the main entrepreneurs here. By steering away from areas with strong national policy traditions (like pollution regulation and taxation), the BEST group managed to create a space for EU policymaking related to carbon regulation in the late 1990s. The group engaged in various types of entrepreneurial activities.

First, the BEST group improved their knowledge position by organizing and utilizing several consultant reports. Here we may also note the lack of feasible EU-wide policy alternatives. In this entrepreneurial effort, the Commission was to some extent assisted by industry actors – for instance, EURELECTRIC carried out emissions trading modelling exercises in 1999 and 2000. Second, within the Commission, while preparing the initial ET Directive proposal in 2001, DG Environment successfully fought back efforts from other DGs such as DG Enterprise to weaken the proposal. In the subsequent EU processes, the clear impression is that the ETS remained something of a DG Environment 'pet project'. Indeed, it seems striking that the ETS became an EU cornerstone based on a very modest administrative foundation – a flagship steered by a minuscule but dedicated crew. Third, an important further element in Commission entrepreneurship was the establishment of a network of stakeholders positive to trading. The formal platform for this was a working group under the first European Climate Change Programme, led by the BEST group, who used agenda setting, participation, and formulation of proceedings as subtle steering instruments. Positive attitudes to market-based ideas among the electricity utilities facilitated this entrepreneurial effort.

The skilled entrepreneurship of the Commission was not sufficient to get stakeholders to support the Single European Market *design*. The 2003 piecemeal markets outcome failed to create a stable carbon price, and hence led to negligible reductions in emissions. A key question then becomes: was entrepreneurial activity equally important in the Commission's later efforts to gain acceptance for changes towards a more centralized and auctioning-based ETS? We would say no – simply because the underlying need for such a tightly-steered process was no longer there. Emissions trading had now taken hold.

This does not mean that a networking dynamic was totally irrelevant in the revision process. A main element was the four meetings within the second European Climate Change Programme's 'working group on ETS reform'. These meetings were forcefully led by BEST entrepreneurs, but participation was broader and less 'steered' this time. Still, some participants at these meetings have mentioned to us that they felt the process was deliberately steered in certain directions. All in all, it seems plausible that the conclusions from these meetings served to underpin and legitimate the quite radical propositions for ETS revision that the Commission put forward in January 2008. Moreover, due to frustrating experiences with the initial ETS, the Commission (subsequently backed by the Parliament) was able to promote Single Market propositions with far greater credibility this time. This can be seen as the fruits of the entrepreneurship of the first phase. By quickly adjusting to the dominant preferences of industry and member states, and concentrating on getting the requested type of system adopted, the Commission managed in a way to have their cake and eat it too: by being recognized as faithful servants to member states *and* by setting in motion a dynamic that could lead up to the system they themselves preferred.

Were there any entrepreneurs who *opposed* centralization and a clearer market approach in the revised ETS? Together with energy-intensive industries, DG Enterprise was instrumental in establishing the High-Level Group on Competitiveness, Energy and the Environment in 2005. This group helped draw attention to the issue of windfall profits reaped by power producers, but it is questionable whether this can be seen as an alternative network to those developed by the market proponents. What may have been lacking was an obvious rallying point for such a network, as there were several possible routes to protect energy-intensive industries. For instance, there was a 'more auctioning and market' route, which would target power producers and reduce their possibilities for windfall profits. But there was also a more 'anti-market' route, involving continued free allowances for the energy-intensive industries. The energy-intensive industries do not seem to have managed to mobilize much support within the European Parliament: the few market sceptics there never tried to establish 'opposition networks' like those seen in the case of renewable energy (see Chapter 5).

To sum up, network entrepreneurship sheds considerable light upon why the ETS was launched in the first place, with the Commission as the main entrepreneur. As to more specific industry entrepreneurship, the most notable example is the 'campaign' of the energy-intensive industries which resulted in the adoption of significant exemptions from the dominant drive towards a Single European Market system from 2013 on. Moreover, and in line with the dominant view in Multi-level Governance (MLG) contributions (see Marks, Hooghe and Blank 1996, Hooghe and Marks 2001), the Commission has been a far more important entrepreneur than industry actors. We have also emphasized here the need to adopt a long-term perspective on the entrepreneurial success of the Commission. Arguably, important reasons why the Commission had considerable success in getting its preferred system adopted in 2009 can be found in clever manoeuvring back in the early 2000s.

So far, we have seen that an entrepreneurial mechanism helps to explain why ETS emerged as a measure and that the persistent entrepreneurship of the Commission eventually bore fruit also with respect to creating an almost 'Single European Market' outcome. However, institutional feedback also plays a major role in explaining why the ETS ended up with this character. As regards industry influence, it took place far more through institutional than entrepreneurial mechanisms.

Policy Interaction: Of Marginal Importance

The ETS was not the first climate policy idea to be discussed in the EU (see Chapter 3). However, these earlier efforts had primarily ended up as weak and symbolic policies with the focus on technology development and very little centralization of control. Eventually, as the ETS rose to become the flagship of EU climate policy, it became increasingly important for ETS promoters to ensure that other EU policies worked well in conjunction with the ETS. Moreover, over time, both the renewable energy policy and the ETS developed in a centralized direction, particularly the ETS.

What does this mean then in terms of *interaction mechanisms?* A precondition for functional interaction to occur is that EU policies, adjacent to the ETS, create tangible effects that again play into the functioning of the ETS or the policy process leading up to ETS decisions. Since the oldest EU climate policy areas – renewable energy and energy efficiency – had resulted in mere symbolic policies, we have not come across such a dynamic. A reasonable interpretation of the parallel increased centralization of renewable energy and ETS is that both policy areas were affected by the growing political importance of climate change, not that the two policy areas affected each other. To some extent the ETS and renewable energy policy target the same main industrial actors, the electricity utilities in particular. But whereas this group achieved a prominent position in the ETS context, it had less success in relation to renewables. The utilities simply did not manage to transfer political standing and clout from the one context to the other. However, their lack of influence on the renewable energy policy did not have any negative effect on their ETS impact. It is important to note that if the ETS had resulted in a higher and more stable carbon price it would probably have affected the realization of renewable energy and energy efficiency to a much greater extent. Hence, in such a scenario, more substantial functional interaction between the policy areas would have been likely.

Concerning *bargained interaction,* we see limited strategic use of issue linkages in the initial ETS process. The original 2003 ETS Directive was developed in a special policy process, with little room for actors to initiate linkages to other areas. This was different when the climate policy package was negotiated towards the end of the decade. Not least, we see some relevant bargaining trade-offs in the final negotiations. But these cases – especially where countries managed to water down the revised ETS by agreeing to stricter policies in other issue-areas – can at best shed some marginal additional light on the 'revolutionary' character of the ETS revision. The transformation of ETS was heavily dominated by issue-internal dynamics.

Was there, then, any *institutional interaction?* Did the design of other policies diffuse to the ETS? Back in 1998, emissions trading was a new instrument in EU environmental policy. But it may still be claimed that there was some important, more overriding, institutional interaction at play, with the ETS more diffusely inspired by the (single) market thinking so central to the EU as a social project.

In the other climate-policy areas in focus in this book we do not see any such relationship: only the ETS stands out as a policy with a strong market character. The other policies were also so weak that they do not seem to have had much effect on the steering method chosen for carbon regulation.

On the whole, we would say that *policy interaction* does not provide any main keys to understanding the 'ETS revolution'. None of the policy interaction mechanisms played any significant role for the ETS outcome. Of course, this situation may change in the future, with the carbon price affected by the performance of renewables and energy efficiency. As we will see in subsequent chapters, what has been far more prominent is the role of the ETS as a sender of policy signals.

The External Environment: Substantially Helping Entrepreneurs in Several Rounds

The emissions trading idea was developed within the global climate regime *before* the EU developed a system of its own. Moreover, important decisions on EU emissions trading policy have correlated with main developments and milestones in the global climate regime. This pattern has occurred at least twice: around the turn of the millennium and in 2007–2009.

Do any of the three perspectives provide us with mechanisms for explaining how and to what extent the EU ETS was affected by outside factors? Let us first look into possible *coercive pressure*. The Kyoto Protocol came to include flexible mechanisms, but it did not formally *require* the development of an EU ETS. Various means and regulatory routes could meet the EU's obligations under the Protocol. Moreover, the US withdrawal from the Protocol in 2001 rendered the regime fragile, with little of a credible market component. Furthermore, related to the lack of development of a global carbon price, we do not see any significant market changes.

On the other hand, we have seen that the increasing attention given to the possibility of global competitive and market effects came to underpin the calls for exemptions to the Single European Market outcome. It was the *failure* of the global climate regime that is the main point here: the failure to develop a comprehensive regime that could introduce reasonably similar carbon constraints to those faced by EU actors in main competitor countries like the USA and China. In such a perspective, LI thinking can sensitize us to factors that impeded the general development of a more streamlined market system. But coercive pressure cannot explain the main changes in the system – towards greater centralization and market streamlining.

What about *institutional diffusion* then? First, it should be noted that the adoption of the flexibility mechanisms in the Kyoto Protocol served as an important stimulus for the EU's subsequent volte-face and the development of an ETS (see Oberthür and Gehring 2006). However, few other parties to the Kyoto Protocol reacted similarly to the EU by developing their own emissions trading systems, so we cannot argue that the world outside the EU exerted strong institutional pressures towards the adoption of a Single European ETS Market. Instead, the EU became a major global proponent of emissions trading as a central climate measure. Hence, it seems that developments within the global regime affected the EU, but the institutional mechanism was not the only mechanism through which the climate regime affected policy development.

What role did *international entrepreneurship* play? We noted above that the BEST group made reference to the flexible mechanisms adopted in the Kyoto Protocol to legitimate the development of an EU system, indicating that an EU ETS from 2005 would 'provide invaluable practical experience of trading'. However, this entrepreneurial element was characterized by other features than those most closely affiliated with the MLG perspective. Let us have a closer look at this.

First, the US withdrawal from the Kyoto Protocol in March 2001 provided an extra opportunity for the EU to play the international carbon leadership card. The Commission entrepreneurs saw this as a 'window of opportunity': they argued that, in order to bring the international process forward, the EU would have to develop significant internal policies. This made it much easier to frame an EU ETS as a core instrument for bolstering EU global leadership. Thereby the perspective of the global climate regime as a 'target' for EU-internal processes was introduced and applied by EU policy entrepreneurs.

On the other hand, as a more enduring, long-term cognitive effect, the lack of comparable climate-policy action in the USA – the key economic competitor to the EU – has acted to impede EU efforts. The US impasse in climate policy has functioned as a legitimating concern for actors within industry, member states and EU institutions, warning of the detrimental effects of an overly strong and front-running EU system that would impose tougher carbon constraints than its competitors. This may in fact be interpreted as a sort of entrepreneurship on the part of industry, cleverly allying with influential forces in key member states, in the Commission and in the Parliament.

Second, as discussed in Chapter 3, the new EU climate and energy targets and policy package put forward in 2007 and 2008 were cast as means and instruments for achieving an ambitious and comprehensive agreement at the 2009 Copenhagen Summit. The keen time pressure and the tripartite process involving meetings between the Commission, Parliament and Council probably led to greater Commission influence on the decision-making process than would otherwise have been the case.

Third, the 'cornerstone' element of the ETS would seem a main thing to note. It was precisely the global regime that legitimized the ETS – and this helps to explain why many actors came to see a centralized, streamlined and smoothly functioning ETS as a sheer necessity. Such an ETS could bolster EU ambitions for global leadership and also strengthen the EU's ability to comply with anticipated, more ambitious post-Kyoto global rules.

Hence, we can conclude that the external environment primarily affected policy development through entrepreneurs that actively referred to external developments, and particularly developments within the global climate regime, both to get the initial system established and subsequently to legitimize their favoured outcome: a Single European Market. The entrepreneurship had a stronger institutional flair than hinted at by MLG perspectives: we suggest calling it *international institutional entrepreneurship*.

Conclusion: The Triumph of 'Tortoise' Entrepreneurship helped by External Events

Why did the EU choose emissions trading as its flagship in regulating carbon emissions? Here, it is important to recall the failed efforts in the early 1990s to develop a carbon tax and an effective common EU climate policy. The possibility of including carbon regulation more strongly in the IPPC regulatory system was briefly discussed and rejected. Other EU climate measures at the time, like the ALTENER renewables and SAVE energy efficiency programmes, were weak, so there was certainly both room and need for a more binding and effective EU measure.

The revised EU ETS that was formally adopted in 2009 went a long way towards establishing a Single European Market for allowances. Why did the ETS finally end up with such a system? We have discussed how and to what extent industry, policy interaction and the external environment affected the outcome. Industry was not very important initially, although it eventually gained more influence over the policy process, and policy interaction was not of high importance in any stage. However, the environment external to the EU did play a crucial role – mainly because Commission officials used developments in the international climate regime actively and skilfully in their entrepreneurial promotion of a Single European Market outcome.

This case shows that entrepreneurs can be highly successful if they adopt a persistent and patient attitude. We could say that the Commission entrepreneurs were not unlike the slow and steady tortoise of Aesop's fable. The Commission officials combined network entrepreneurship and international institutional entrepreneurship. The successful network entrepreneurship exercised by the Commission in the 1997–2004 period was instrumental in setting in motion and shaping the 'revolution' that ended in the adoption of a far more Single European Market-oriented system in 2009. In the first period, the Commission adjusted quickly to the dominant preferences of industry and member states, and concentrated simply on getting a market-based system adopted, rather than a perfect market-streamlined and highly centralized system. In this way, Commission officials managed to gain recognition as faithful servants to member states *and* set in motion a dynamic that could lead up to the system they themselves preferred.

In the last period, after 2005, the Commission was able to reap the profits from their initial modest approach. Now they could argue that further streamlining and centralization was needed in order to improve the functioning of the system.

Moreover, the first ETS had created institutional feedback effects that underpinned these arguments. In addition, the Commission repeatedly referred to international climate arguments for a Single European Market outcome.

Two additional mechanisms provide valuable additions to this main explanation. First, the *instrumental governments* mechanism, a mechanism exerted from the LI perspective, must be included in the explanatory picture. ETS promoters in Brussels as well as at the national levels all along enjoyed considerable understanding and support from the main industrial actors targeted by the ETS – the utilities. In addition, the energy-intensive industries managed to get some national governments to act as their instruments, more hampering than underpinning the development towards a centralized, market-streamlined system. Second, also *policy interaction* contributed to water down the Single European Market flair of the outcome – in particular, member-state-induced bargained interaction.

Hence, both the instrumental governments mechanism and the bargained interaction mechanism contributed to induce exceptions to a fully-fledged Single European Market outcome. However, it was the tortoise entrepreneurship of the Commission – based on repeated use of network entrepreneurship and international institutional entrepreneurship – that provided the main 'revolutionary' driving force in this case. That the subsequent financial crisis and developments indicate that the 'institutional surgery' was only partially successful is another story, to be told in coming publications.

Chapter 5

EU Renewable Energy Policy: David beating Goliath?

Introduction

The EU started to develop a policy on renewable energy back in the 1970s, but it was not until 40 years later that a forceful EU policy came into being. Directive 2009/28/EC (European Parliament and Council 2009a) on renewable energy, similar in character to the national industry development projects of the postwar decades, aims at achieving a 20 per cent renewables share by 2020. This will require a significant increase, as only 8 per cent of the energy consumed within the European Community in 2007 stemmed from renewable sources (European Commission 2010b). That 8 per cent came chiefly from old hydropower plants and the development of solar power and wind energy in Denmark, Germany and Spain. With the 2009 Renewables Directive, all EU member states must step up their involvement. This chapter examines how the 'EU engineering' renewables policy came about.

According to conventional wisdom in political science, policy outcomes will tend to reflect the interests of economically powerful actors. This line of argument is also strong in European integration theory, not least within the Liberal Intergovernmentalist school, which holds that member states will act as instruments for dominant national industries. The EU policy on renewable energy emerged through heated conflict between economically powerful electricity utilities and actors in the much smaller renewable energy industry – the former preferring a Single European Market policy and the latter opting for what we call 'EU engineering' types of policies.

Quite surprisingly, it proved to be the economically weakest group – the renewables industry – that had greatest influence in shaping the policy outcome. How was it possible to dwarf the power of the big utilities? How did David manage to beat Goliath? This puzzling outcome makes the case of renewable energy instructive for exploring how the power of industry actors plays into EU policymaking.

Moreover, we note that the major EU decisions on renewable energy policy have correlated with main decisions in the global climate regime. This pattern has been repeated three times: in the early 1990s, at the entrance to the new millennium, and in 2007–2009. Although renewable energy is a common low-carbon technology option, it is still difficult to trace a direct relationship between global developments and EU policymaking. After all, there have been no global decisions that would require the EU to develop renewable energy, so the EU could just as well have opted for other climate solutions. Why then this clear pattern between global and EU-level developments?

On this backdrop, the two main questions addressed in this chapter are: Why has the EU developed a policy on renewable energy? And why has this policy ended up with an EU engineering character? We begin by exploring how industry affected the policy outcome. Next, we investigate whether interaction with other policy areas contributed to the EU engineering outcome. Third, we ask: did external developments play into EU decision-making? In line with the discussion in Chapter 2 in this book, Liberal Intergovernmentalism (LI), New Institutionalism (NI) and Multi-level Governance (MLG) guide our search for causal mechanisms.

The Renewables Policy Outcome: EU Engineering

Directive 2009/28/EC,[1] the renewable energy directive (European Parliament and Council 2009a) is the core of the EU's policy on renewable energy. It now provides Brussels with substantial authority as well as more incentives for technology development than the use of market measures.

Concerning the steering method, the 'technology development versus market' dimension, the directive clearly leans towards technology development. All member states are obliged to reach national renewable share targets; further, 'special attention can be given to sectors that suffer disproportionately from the absence of technological progress and economies of scale and therefore remain under-developed, but which, in the future, could significantly contribute to reaching the targets for 2020' (Preamble 20). Each member state is to report to the Commission on how it has 'structured its support scheme to take into account renewable energy applications that give additional benefits in relation to other comparable applications, but may have higher costs' (Article 22). Clearly, then, the directive encourages member states to support the development of high-cost technology.

The Renewables Directive introduces three flexible mechanisms, designed for 'facilitating cross-border support of energy from renewable sources without affecting national support schemes' (Preamble 25). First, statistical transfer mechanisms enable member states to make arrangements for the statistic transfer to other member states, so that renewable energy produced in one country can be counted towards the target of another (Article 6). Second, the joint projects mechanism serves the same purpose, but involves collaboration during the whole planning process (Article 7). Third, the joint support scheme mechanism enables member states to 'decide, on a voluntary basis, to join or partly coordinate their national support schemes. In such cases, a certain amount of energy from renewable sources produced in the territory of one participating Member State may count towards the national overall target of another participating Member State' (Article 11).

---

1   Also referred to as 'the Renewables Directive' and 'the Renewable Energy Directive'.

These mechanisms give member states some flexibility with respect to achieving their targets and enable the development of cross-national schemes, whether market-based or not. As we shall see, certain actors have promoted the development of 'green certificate' schemes that oblige energy producers to offer their customers a certain quota of renewable energy and create a market of securities that the energy producers can buy if they do not provide the renewable energy quota themselves (see European Commission 2008j). However, the Renewables Directive does not *require* member states to develop such green certificate schemes or other market measures. The three mechanisms may facilitate various kinds of cross-border collaboration – but whether this underpins the development of a pan-European green certificate scheme or technology-development steering methods will depend on how the member states employ these mechanisms. By spring 2012, very few countries have set about using the mechanisms, so it seems unlikely that they will affect the steering method of the EU policy on renewable energy much (Ragwitz et al. 2011).

We then turn to the centralization dimension. According to Article 3 of Directive 2009/28/EC (European Parliament and Council 2009a), the aim is to increase the overall share of energy from renewable sources in gross final consumption of energy to 20 per cent by 2020. Various types of renewable energy, in the form of electricity or heat applied for stationary purposes or for transportation, count towards fulfilment of the target. With regard to transport, '[e]ach member state shall ensure that the share of energy from renewable energy sources in all forms of transport in 2020 is at least 10% of the final consumption of energy in transport in that member state' (Article 3). If the overall volume of energy consumption is reduced, the size of the renewable energy percentage will increase. This implies that energy-efficiency improvements will contribute to achievement of the target. Most importantly, mandatory national targets are set for all member states. These individual targets are developed on the basis of the economic strength of each member state, not its technological potential for producing renewable energy. The various national renewable energy action plans show that member states have planned a significant increase in energy efficiency, and are heading for a shift in the annual increase in production of renewable energy from 22 TWh a year in the period 2000 to 2008 to 52 TWh per year for the period 2008 to 2020 (EURELECTRIC 2011: 19).

The Directive gives the Commission substantial possibilities to control and redirect the national policies. Member states are instructed to develop 'an indicative trajectory tracing a path towards the achievement of their final mandatory targets' (Preamble 18). There is a detailed description of how the national renewables targets are to be measured; each member state is also obliged to adopt a national renewable energy action plan and report to the Commission every second year (Article 4, 5). In June 2009 the Commission issued a highly detailed template for reporting, dealing with all kinds of administrative procedures, support measures, information campaigns and the like (European Commission 2009b). The national action plans are to be evaluated by the Commission, which thereby may interfere in almost every detail relating to a member state's national policy on renewables (Article 4). However, it is up to the member states to decide how and to what extent they will exploit the flexibility opportunity offered by the mechanisms.

On this basis we may conclude that the renewable energy policy of the EU has a distinct 'EU engineering' quality: it centralizes substantial authority to EU organizations and relies on technology-development criteria, not economic criteria and market mechanisms.

The Renewables Policy Story: From Mere Symbolism to Powerful Industry Creation

Pre 1997: Emergence of Three Different Policy Approaches

The oil shock in the early 1970s led many European countries to invest in renewable energy research and development (R&D) (Hildingsson, Stripple and Jordan 2010). In the early 1980s, the EU prudently started to promote renewables, through symbolic statements and miniscule financial support. The 1990 Council decision to stabilize CO_2 emissions at the 1990 levels by the year 2000 stepped up this engagement. Aiming to establish EU as a leader in the global environmental talks at the 1992 Rio Summit, the Commission proposed binding renewable targets; a general framework for licensing of small hydro; a directive on the safety, performance and environmental impact of wind turbines; and a directive on the technical specifications of biofuels for diesel engines. In addition, the ALTENER programme was requested to offer R&D support (European Commission 1992a, see also Chapter 3). The Parliament supported these ideas, but the member states rejected them all, except ALTENER – which was realized with low levels of financial support (Hildingsson, Stripple and Jordan 2010: 108).

European states developed three different renewables strategies in the following years. First, Germany, Denmark and eventually Spain launched feed-in schemes, relying on politically decided electricity prices, differentiated for various technologies. This led to the emergence of small-scale renewable energy industries. In Germany, an alliance of engineering research communities, small-scale cooperative ventures and the Green Party (and eventually the Social Democrats) took the lead (Meyer 2003, Jacobsson and Lauber 2006). This alliance ensured a feed-in scheme, offering reliable support for various technologies, with no incentives towards development of the most profitable projects (Meyer 2003: 671, Reiche and Bechberger 2004: 248). Initially, German utilities were not entitled to receive any benefits. Together with the Ministry of Economics, the utilities called for a market scheme that would favour the development of the least costly renewable technologies (Jacobsson and Lauber 2006). In 1996, the utilities lodged a complaint to the Commission. DG Competition (DG Comp) responded by arguing that the feed-in rates should be reduced. This led to massive protests from the renewables industry, and eventually the feed-in regime was strengthened (Jacobsson and Lauber 2006).

Denmark's new renewable energy industry emerged from academic technological communities, environmental groups and farmer communities. The utilities first opposed the feed-in scheme, but later established a wind-power agreement with the Danish government (Meyer 2004: 30). In the late 1990s, Spain introduced a feed-in scheme that rapidly led to major developments in wind power (Reiche and Bechberger 2004: 247, Río and Unruh 2007: 1503). A modest national renewable energy industry emerged, but the large utility Iberdrola took the lead in investments.

In a second group of countries, the traditional electricity producers were central to the development of renewables. Finland, the Netherlands and Sweden strengthened R&D support and other measures, and did not develop feed-in schemes (Meyer 2003, Jacobsson and Bergek 2004). The utilities took the lead in renewable energy development, and no substantial small-scale renewable energy industry emerged. In Sweden and Finland, renewable district heating capacity boomed. The UK can also be placed in this group, although it did not offer much investment support. Instead, it launched a competitive bidding scheme in 1990 (Mitchell and Connor 2004: 1936). Although planning activities increased substantially, few renewable energy plants were actually realized. Also in the UK, it was mainly the utilities that became involved in renewables.

The third and largest group of European countries hardly engaged in renewable energy engagement at all. True, France, Italy, Belgium and Austria developed some policy measures, but ambitions were low and the results meagre (European Commission 1992a, Reiche and Bechberger 2004). In the Eastern European countries (not members of the EU at the time) renewable energy was hardly an issue at all. Most of these countries, including Poland, had energy systems with excess capacities and thus no need for enhanced capacity.

Thus, three distinctive renewables approaches emerged. Denmark and Germany offered feed-in support that led to the emergence of a new renewable energy industry; Finland, the Netherlands, Sweden and the UK developed different schemes, all favouring traditional utilities; and little or nothing happened in the third and largest group of countries. None of the three approaches gave much attention to market thinking. The EU offered modest renewables R&D support but was not granted competence to interfere to any great extent in the development of national policy.

1997–2004: Market Ideas Challenge Technology Development

In the run-up to Kyoto, the Commission stepped up its focus on renewable energy as a solution to climate change and the EU's increasing energy dependence (European Commission 1996, 1997). Again, the Commission underscored the need for a common EU policy on renewable energy, and proposed an indicative target of 12 per cent renewable energy share of gross inland energy consumption by 2010. This was immediately given a mixed reception from member-state energy ministers; and the association of European utilities, EURELECTRIC, voiced scepticism (*ENDS* 1997a, 1997c). On the other hand, various European renewable energy industries supported the initiative, and environmental organizations argued that it would give 'a big boost' to the upcoming Kyoto negotiations (*ENDS* 1997d). The European Parliament called for higher ambitions, but was not heard (*ENDS* 1997b).

The Commission also called for a market approach, issuing a white book stating that liberalization 'can form the basis for a dynamic and secure role for renewables so long as adequate market-based instruments are provided' (European Commission 1997: 15). Furthermore, it argued that common EU rules were becoming 'increasingly necessary as Europe's electricity market is liberalised', adding that if a situation with a range of support schemes across Europe were allowed to continue, it 'appears likely to result in distortion of trade and competition' (*ENDS* 1998a).

Because the EU had taken on a Kyoto obligation on behalf of all its member states, the Commission was given a prominent role in the implementation process (see Chapter 3). The Commission was eager to follow up the Kyoto commitment with the development of a tangible EU policy on renewable energy (*ENDS* 1998a). The Energy Council supported the Commission's emphasis on renewable energy as a key to reach the Kyoto target (*ENDS* 1998b). The European Parliament was impatient and called for member states to set binding national targets to increase the market share of each type of renewable energy (*ENDS* 1998c).

On this backdrop, the Commission drafted a directive requiring member states to produce at least 5 per cent of their electricity from renewable sources. This was not a high figure for the frontrunners, but was challenging for countries with very low renewable shares (*ENDS* 1998e). Further, the Commission sought to steer and limit how the member states could support renewable energy, indicating that feed-in schemes would be allowed only for a transition period, and that all schemes should be mutually acceptable in all EU countries. Thus it envisaged a common European market-based support scheme that could eventually lead to harmonization of national schemes. Our interviews show that Commission officials from various DGs supported this, whereas DG Energy officials primarily preferred a technology-development approach and feed-in schemes. Renewable energy actors supported the development of national targets, but Germany's wind-power industry and the German government protested vigorously against harmonization and marketization. This eventually led the Commission to shelve the draft directive altogether (*ENDS* 1999a).

At this point, the renewable energy industry and renewables activists joined forces in Brussels in order to promote a binding EU policy without incentives towards support-scheme harmonization. National niche organizations relating to wind power, biomass etc. that were already organized in Brussels-based associations joined under the common umbrella of the European Renewable Energy Council (EREC). In addition, Spanish MEPs initiated the European Forum for Renewable Energy Sources (EUFORES), initially having renewable energy companies as well as MEPs as members (EUFORES 2010). Eventually, EUFORES became a more streamlined organization exclusively for parliamentarians. All these actors advocated greater EU involvement in renewables, but initially there was internal disagreement as to which EU steering method was better: technology development that fostered national feed-in schemes, or more market-based approaches (*ENDS* 1999a).

In parallel, the electricity industry started to support the emergence of a pan-European, market-based support scheme. Due to lack of experience with such systems, there was considerable uncertainty as to what such a scheme should or could look like. German utilities, being among Europe's largest electricity producers, strongly supported the development of a green certificate scheme (Rowlands 2005, Foquet and Johansson 2008). Also EURELECTRIC praised this idea. The Commission encouraged the increased engagement of the utilities by way of sponsoring several test and research projects on how a European green certificate could or should be designed (Meyer 2003: 670). The Commission also supported the creation of Renewable Energy Certificate System (RECS), an organization for the promotion of a pan-European market scheme (*ENDS* 2000a). However, with the exception of Spanish Iberdrola and Swedish Vattenfall, few utilities invested much in renewable energy: most continued to rely predominantly on fossil fuels and nuclear power.

The two industry approaches to renewable support reflected differing developments within two groups of member states. Market ideas gained hold in countries where the utilities had the prime role in renewable energy development in the first era, particularly in the Netherlands, Sweden and the UK. In 1998, the Dutch government introduced a voluntary green certificate scheme, exempting from an environmental tax those members of the public who purchased 'green' electricity (Meyer 2003: 672). This proved an instant success, but new demand grew too large to be fulfilled by domestic production, and the Dutch utilities had to buy renewables certificates abroad (Meyer 2003: 672, Reiche and Bechberger 2004: 248). Since that did not foster any actual domestic energy production, the Dutch government then abandoned the scheme, and failed to launch a new system. Sweden launched a national green certificate scheme in 2003 (Riksdagen 2002/03), which immediately boosted investments (Jacobsson and Bergek 2004). The UK designed a quota system, but it failed to result in much production (Mitchell and Connor 2004). Thus, the market idea provided success in only one of three market frontrunner countries.

What happened in the group of feed-in countries – Denmark, Germany and Spain? In Germany, market measures were promoted by the utilities, the Ministry of Economics and Technology (former Ministry of Economics) and conservative politicians, but opposed by the renewable industry as well as the ruling Social Democrat/Green Party coalition government. The German utility Preussen-Electra (later merged into E.ON) brought a case to the European Court of Justice, claiming that the feed-in scheme conflicted with EU regulations on state aid (Meyer 2003). In 2001, the Court dismissed the claim and thus the German feed-in scheme was saved (Kuhn 2001, Jacobsson and Lauber 2006). Moreover, Germany's Greens succeeded in transferring responsibility for renewable energy from the Ministry of Economics and Technology to the feed-in supportive Ministry of Environment (*ENDS* 2005d). These events underpinned enduring German support of national feed-in schemes.

Spain rejected the certificate idea immediately and continued to develop its feed-in scheme, without much political fuss or controversy. In Denmark, however, we can note a radical shift in political attitudes. In the early 2000s, it was decided to replace the feed-in scheme with a green certificate scheme, but this led to turmoil and a substantial drop in domestic investments. The idea was eventually dropped (Meyer 2004, Lov om fremme av vedvarende energi 2008).

At the EU level, the environmental movement argued that the EU's Kyoto commitment would be 'as good as dead' if no renewable energy directive were adopted (*ENDS* 1999b). Eventually, the EU energy ministers stated that they would accept the development of a directive as long as it explicitly permitted national governments to steer the development of national support schemes (*ENDS* 1999c). The Commission started drafting the directive, aiming for binding national targets and a certain degree of market harmonization (*ENDS* 1999c, 1999d). The European Parliament, particularly EUFORES members, repeatedly called for a more ambitious and binding directive (*ENDS* 2000b).

The Commission tried to develop a national target distribution that the member states would see as both fair and ambitious (*ENDS* 1999d). It did not succeed, and ended up proposing a directive with indicative targets, amounting to an increase from 14 per cent to 22 per cent of EU primary energy supply from renewable energy sources by 2010 (European Commission 2000d). Furthermore, the Commission proposed that certification (guarantees of origin) be required of all renewable energy production. The rationale was that such certificates could become tradable securities in the future that would ease the transition to market-based support schemes (*ENDS* 2000c). In addition, it proposed the harmonization of rules for national schemes within five years, as well as procedures for detailed EU monitoring of national progress in renewable energy policy and development. The Parliament called for binding targets, but rejected a deadline for harmonization of schemes (European Parliament 2000, 2001). The draft was met with mixed responses from member states; eventually the Energy Council accepted most of the Commission's proposals, although it watered down the monitoring mechanism (*ENDS* 2000g).

In parallel, the Commission challenged both the German and the Spanish feed-in schemes, claiming they were 'potentially disruptive to the EU's single market' (*ENDS* 2000f). As mentioned, these arguments were repudiated by the Preussen-Electra ruling of the European Court of Justice, when it in November 2000 ruled that Germany's feed-in law did not constitute state aid (*ENDS* 2001b, Kuhn 2001). As a result, the Commission could not prevent the feed-in scheme idea from spreading.

Directive 2001/77/EC (European Parliament and Council 2001) was finally adopted in September 2001. A few countries had negotiated down their indicative targets somewhat, but the total figures remained largely unchanged. The Directive contained no direct market or harmonization pressure, but it required member states to issue guarantees of origin – a statement declaring whether the electricity was from renewable sources. This was a technical procedure on which a future mandatory scheme could be based. Introduction of this procedure enabled the electricity utilities to offer securities to customers who would pay a little extra for green electricity. In this way the consumers contributed financially to the production of renewable energy, but they were not guaranteed that they themselves would be supplied with the electricity generated by a particular source or plant. In addition, in 2005 the Commission was tasked with reviewing the functioning of the existing support schemes, in order to put forward recommendations as to further harmonization and marketization. All in all, the Directive had what we would call a 'local loading' quality: it promoted increased national efforts, but offered few incentives for the development of national market instruments.

What happened after the 2001 Directive was adopted? Most importantly, only a handful of countries developed national green certificate schemes, and the voluntary certificate scheme (based on guaranties of origin) largely failed (European Commission 2008j, Foquet and Johansson 2008). Hydropower companies used the voluntary system as a way to increase their profits from old hydropower generators, but this was controversial and did not help to raise the legitimacy of market measures. Second, the Commission drew the conclusion that the narrow focus on electricity was a mistake, and that renewable district heating was needed in order to increase the share of renewables. Third, the accession of eight Eastern European countries in 2004 increased the number of EU member states that had no national renewable energy policies (Reiche 2006).

Thus, we may conclude that as of 2005, the member states had still the upper hand in EU renewables policy. The feed-in countries managed to spur small but prosperous renewable industries, while countries that applied market measures had less tangible results to show. Moreover, the new renewable energy industry had developed a strong presence in Brussels.

2005–2010: Increased Conflicts and Stronger Renewable Industry Dominance

From about 2005 and onwards, renewable energy became a more pressing issue, for industry and politicians alike. The Commission argued that the current global situation warranted the development of a more ambitious EU directive on renewables (*ENDS* 2007a). This was supported by core member states like Germany and the UK, both of which aspired to act as leaders in the international negotiations (Rayner and Jordan 2011, Jänicke 2011). Renewable energy niche organizations, gathered in EREC, exploited this window of opportunity, demanding a more ambitious and comprehensive EU policy on renewables, and campaigning for a heating and cooling directive as well. EREC calculated that the renewables industry could deliver a 20 per cent renewable energy share by 2020 (EREC 2004). These claims were supported by the European Parliament, the EUFORES members in particular (European Parliament 2006).

In 2005, the Commission issued a report, in line with the decision in the 2001 Directive (European Commission 2005a). The report was a compromise between the technology-development supporters in DG Energy and the market proponents in other DGs. It concluded that 'the harmonization of support schemes remain a long term goal (...) [but] harmonization in the short term is not appropriate' (European Commission 2005a: 17). The Commission started to revise the renewable energy directive, and soon concluded that it would be easier to gain support for one binding target for all energy sources and sectors, than to develop different binding targets for electricity, for heating, for cooling, and for transport. A Commission interviewee explained: 'It was primarily because we wanted to give the countries more freedom that we decided to include all sectors in one directive (...) This allowed us to steer more rigidly on an aggregate level.' Our interviews show that this approach soon obtained consensus among most actors involved in the deliberations.

In the run-up to the European Council meeting in March 2007, the renewables industry lobbied intensively for a binding renewables target, but our interviews show that they had little real hope of such an outcome. Thus it came as a surprise when the heads of state actually decided to go for a binding renewables target (European Council 2007). Some interviewees give the Commission credit for this; others see the main drivers in the high climate ambitions of Merkel and Blair, with Merkel in the forefront as Germany had the presidency at the time. The latter view has been supported by studies of UK and German engagement in EU climate policy, showing that both countries sought to play key roles in European as well as global negotiations (Rayner and Jordan 2011, Jänicke 2011). As discussed in Chapter 3, the 2007 Council decision was made against the backdrop of slow negotiation progress in the global climate regime and the EU's worsening energy security situation.

At this stage, the target was not broken down into sub-targets for each member state and the Commission had not developed proposals for measures. Immediately after the 2007 Council decision, DG Energy informally proposed national targets, based on the countries' differing theoretical potential for producing renewable energy. The assumption was that it was not important exactly where the renewable energy was produced, as long as production involved 'the lowest possible costs'. This approach did not work out. According to one Commission interviewee: 'We started to test this on ministers from the member states. They reacted fiercely. Some were actually frothing! (...) One went totally red'. Countries that already had large shares of renewable energy and those with scarce financial resources were particularly upset.

This failure gave new room for the market promoters in DG Energy, some of whom had been the architects of the EU ETS a few years earlier. These officials readily developed a key for target distribution, based on a 5.5 per cent flat target for all countries and an additional percentage to be calculated on the basis of the countries' GDP. Application of this formula obliged all countries to enhance their share of renewable energy consumption, but the level of ambition was related to their economic strength. Moreover, the new approach did not require the domestic renewable energy share to stem from national production. The plan was to combine this with the introduction of a pan-European certificate scheme, allowing member states to invest in renewable energy in countries with better technical potential. Our interviews confirm that the new proposal received favourable responses from member states. It was perceived as balanced and fair, and only minor changes were made in the process leading up to the Commission's launching of the draft directive in January 2008. A key process insider explains: 'It was a zero-sum game: if some claimed that their target ought to be reduced, then some others would have to increase their target.'

The introduction of binding national targets radically strengthened centralized control, but all this was smooth sailing compared to the discussions on steering method. By 2008, only seven EU member states had green certificate schemes, whereas 18 had adopted feed-in schemes (European Commission 2008j). Despite the modest diffusion of the market idea, the utilities and certain Commission officials again tried to introduce a pan-European green certificate scheme. The Commission also concluded that because the ETS created a carbon prize that was too low to spur the development of renewable energy, additional EU measures were needed (European Commission 2008k: 11). In the course of 2007, market supporters in the Commission produced a rough draft of a full-fledged EU certificate market that would shift competence from member states to EU organizations, and over time probably lead to the elimination of existing national feed-in schemes (European Commission 2007c). Support would be granted only to the least costly renewables projects, and governments would no longer be able to favour specific technologies and power generation at specific geographical localizations.

In order to understand the responses to this Commission proposal we need a deeper understanding of the two distinct organizational fields, with different social orders, that had emerged: the field of European electricity production and the field of renewable energy. Let us first explore the electricity field. Chapter 4 showed that the electricity producers had increasingly taken up market thinking, as well as Europeanizing their activities. At first, the seven dominant utilities had ensured growth by mergers and acquisitions. A conglomerate of regional European electricity markets had emerged – neither unified nor uniform (Glachant 2003, Rademaekers, Slingenberg and Morsy 2008: 18–19). Poor transmission possibilities represented severe obstacles to trade, and the incumbents were operating under a level of national market control that severely hampered the functioning of market forces. Because EU competition law to some extent constrained the majors from growing more in their domestic markets, the predominant strategy became acquiring shares in new markets, often through swap deals (Codognet et al. 2003, Glachant 2003). However, by 2008 the opportunities for applying this strategy were limited by lack of potential acquisition candidates in Europe (Vattenfall 2009: 30).

In this situation, investment in renewable energy became one of the few options that could allow the dominant utilities to grow (see EURELECTRIC 2005, Vattenfall 2009: 30). Only Vattenfall and Iberdrola had historical roots in renewables, in the form of large-scale hydro (EDF Group 2009, E.ON 2009, ENEL 2009, GDF Suez

2009, Iberdrola 2012, RWE 2009, Vattenfall 2009). The two had also developed strong new renewables portfolios – Vattenfall in biomass and Iberdrola in wind power. A radical shift took place in 2008: now all companies created subsidiaries or business units for renewable energy and acquired renewable energy companies or renewable power plants. Further, they expressed more or less specific ambitions as to growth in renewables investment. Our business interviewees report that this also resulted from the increasingly radical signals from the EU as well as national governments. As one industry interviewee put it: 'now that the governments talk about renewable taking over the market (....) we have to reconsider.'

Because so much was at stake, the utilities engaged more in a campaign for development of a pan-European green certificate scheme than ten years earlier. There was one main exception, however: the large Spanish utility Iberdrola, which opposed green certificates and defended national feed-in schemes instead. Our interviews show that the market approach attracted considerable Commission support, from DG Competition and DG Enterprise, but also from prominent DG Environment and DG Energy officials. Why then was the outcome still so meagre? For answer that, we must explore how the renewable energy actors countered the pro-market initiatives.

Striking and significant changes within the European renewables industry had played out during the 2000s. As described by one interviewee: 'I remember 10 or 15 years ago (...) when I went to renewables meetings they were dominated by old men – the inventors – and the young environmentalists. It was a strange crowd. Now it is dominated by big business.' However, despite record-high annual growth rates, the renewables industry could still not compete with the utilities in economic strength. In 2008, only a handful of renewable energy companies made it to the list of Europe's top 500 companies, and none figured in the first part of the list (*Financial Times* 2008a). Germany was the only country with a significant renewables industry, although Danish wind-power companies were also significant in size (Meyer 2003, Jacobsson and Lauber 2006).

Furthermore, the renewable energy industry had developed an exceptionally strong Brussels community, with varying associations co-located in 'the Renewables House'. The umbrella organization EREC coordinated all niche organizations. Our interviews show that close links existed between the industry, the technology-development supporters in DG Energy (named DG Tren at the time) and various European Parliament groups. In addition, industry actors had strong ties to the Danish and German governments, but also to the Spanish government despite the lack of a strong renewables industry there. Industry, Commission officials, member-state representatives as well as MEPs had become more united in their scepticism towards market harmonization and feed-in support. They now shared the view that a pan-European green certificate scheme would have unpredictable effects on the development of renewable energy, and that the utilities would profit unreasonably. Moreover, the renewables actors had strengthened their lobbying capacity significantly: according to interviews, the staff of the EREC organizations was three times bigger in 2010 than it had been only five years before.

The conflict between the organizational field of European power production and the renewables field was not a mere clash of economic interest, but also a conflict between two institutional logics: market versus technology development logics. Our interviewees describe the conflict as having an 'almost religious' character, and the media coverage provides many examples of actors accusing each other of fraud, lack of credibility, being reactionary, and so forth (see *ENDS* 2005a, 2005c, 2006). There were conflicts within national governments, in national media, within the Commission, and between the two energy industry groups.

Let us now return to the chronological story. According to interviewees, DG Energy officials opposed to the market approach ensured that the Commission draft directive was 'leaked' in December 2007 (European Commission 2007c.). This happened only weeks before the Commission was to launch the full climate package. Hence time was in short supply when the EREC organizations, the German and Spanish governments, and Iberdrola engaged in intense lobbying of the Commission. According to interviewees, market proponents like EURELECTRIC and the UK did not engage in similar entrepreneurial efforts to any great extent.

One month later, the Commission issued a new and rather inconsistent draft directive, opening up for certificate trade but not creating a pan-European scheme (European Commission 2008d). The renewables actors were still not satisfied, and our interviewee explain that they argued that, because certificates were defined as 'tradable goods', they would become subject to EU competition legislation. This could enable the European Court of Justice to rule that member states should include all applicants in their support schemes, even if the plants were constructed outside the country. That would attract too many applicants to the most generous feed-in schemes, like the German scheme, and would in turn lead to breakdowns. Interviews show that this argument eventually hit also the Commission trading promoters.

Claude Turmes, vice-president of EUFORES (later president) and Green Party representative, became the Parliament rapporteur for this directive. As noted by our interviewees, he created a group of German and Spanish civil servants, EREC representatives, and some researchers to facilitate drafting of the Parliament's report. This group aimed to secure for countries with scarce renewables potential the possibility to invest in renewables projects elsewhere, without creating an EU-wide market scheme. One interviewee has described the process:

> At first we tried to develop a system with flexibility, but gradually it just became more and more complex and we stopped believing in it. We understood that it was impossible to develop a scheme that could function. Yet we launched it. And that was due to tactics. We wanted to show how complicated this was.

This tactic was certainly successful. Turmes created a solid alliance in the Parliament that included all the major groups, except the conservative group PPE (*ENDS* 2008h). Many interviewees emphasize that Turmes exercised great entrepreneurial skills during this process, with regard to his ability to create wide

Parliamentary support, but also his engagement in the wider European debate. In conjunction, EREC representatives and Turmes approached the member states, seeking to build wide support for his proposal. They did not criticize the market idea in bold terms, but instead argued that it was risky and uncertain, whereas the feed-in schemes were already delivering tangible results. Although much of the controversy centred on the steering method, it is worth noting that the Parliament also introduced stronger centralization of control than proposed by the Commission. By developing a detailed action-plan template the Parliament succeeded in introducing formulations in the proposed directive that allowed the Commission to control how member states developed their national policies on renewable energy (European Parliament 2008b).

The launching of a joint compromise proposal from the UK, Poland and Germany in June 2008 was a major breakthrough for the tactic of the renewables actors (Non-paper 2007, *ENDS* 2008f). The proposal clearly stated that member states should remain in control as regards designing their national support schemes. The non-paper proposed three flexible mechanisms: a joint support scheme, statistical transfer, and joint projects. The mechanisms would primarily allow bilateral collaboration, but it would be up to member states to decide whether to apply these mechanisms. The Commission was not granted any role with respect to the design of national schemes. Interviewees agree that this changed the dynamics of the deliberations over the directive. As one put it: 'When these three countries had decided to all be in bed with each other it was hard to redirect the process. It was unstoppable!'

The compromise represented a major shift in the UK position: after having been the main market promoter for a decade, it now basically gave this idea the kiss of death. Interviews indicate that the main reason was that the UK had been striving to make its own market system work. We may also note that, due to the absence of strong British utilities, UK policymakers were not exposed to strong promoters of national green certificates. British renewable energy enthusiasts had repeatedly called for the introduction of feed-in systems, due to the disappointing achievements of the market scheme (Mitchell and Connor 2004). This appears to have led to a change in the position of the UK government: for instance it planned to introduce a string of new, domestic non-market measures, even a national feed-in scheme to supplement its market scheme (effectuated after the final decisions on the renewables directive) (HM Government 2009).

Interviews indicate that Poland joined this initiative largely for tactical reasons. By appearing constructive in this issue, it could improve its chances of getting special treatment for its coal-based utilities in the ETS, Poland's main policy priority at the time (see also Jankowska 2011). Note that the severe time-pressure (due to the objective of reaching an agreement that could move the global climate talks, see Chapter 3) created extra pressure on actors to create viable compromises. Interviewees indicate that this made it easier for renewable energy entrepreneurs to influence the UK and Poland.

By now, the market proponents found themselves fighting an uphill battle. As remarked by one civil society interviewee, 'the Swedish government was the only one that was a little bit sympathetic towards our viewpoints. The UK stopped to give priority to this after they supported the compromise.' When asked why utilities did not achieve a stronger impact on the policy outcome, a utility representative sighed: 'we are simply not politically correct'. Making a comparison with the ETS process, an EU official stated: 'It is so much easier when there is nothing to build on. But here there were already national systems in place. That makes it much more complicated'.

As chronicled in Chapter 3, the deliberations ended in a trialogue process with representatives from the Commission, the Parliament (Turmes), and the French Presidency. Interviewees agree that most issues were resolved before the final rounds of the trialogue. Hence, the renewables directive was not particularly involved as a bargaining chip in the concluding negotiations concerning the entire package.

We can conclude that the conflict over the steering methods for renewable energy proved to be one of the most heated conflicts during deliberations over the EU climate package, while increased centralization of control followed along a smoother path. Renewable energy actors had the upper hand in the last phase of policy development. However, adoption of the new renewable directive failed to create a viable compromise with respect to support schemes. When the member states had submitted their national renewable energy plans, showing that that they would primarily use technology-development measures and hardly apply the flexible mechanisms, the Commission responded by calling on member states to boost cooperation and to design cost-efficient support schemes (European Commission 2011a).

Energy Commissioner Günther Oettinger said: 'We have to invest much more in renewable energy and we need smart, cost-effective financing. If Member States work together and produce renewable energy where it costs less, companies and consumers and the tax payer will benefit from this' (Europa 2011). EU officials have stated in interviews that they were preparing for new rounds on the steering method: 'We also must take into account what has happened since the directive was decided upon. Spain, the Czechs and even Germany have cut their support schemes. They simply cost too much!'

The renewable energy industry has argued that Europe has the technical potential for an even more rapid increase in renewable energy production than the planned shift from 22 TWh a year in the period 2000 to 2008 to 52 TWh per year by 2008–2020 (EURELECTRIC 2011: 19). As for the electricity industry, it has focused on the technical and economic challenges related to this shift. Although they no longer use the term 'green certificate', they still call for the 'establishment of a European level playing field' and 'progressive implementation of an EU-wide harmonized RES support system which is compatible with a common European electricity market' (EURELECTRIC 2011). Civil society interviewees have indicated that renewable energy has become increasingly important for the utilities, particularly after the German government decided to close down much of the country's nuclear capacity. Hence, it seems as if the conflict over steering method is likely to rumble on and on in the EU debate on renewable energy.

But why is it that EU policy on renewable energy has an 'EU engineering' character? Here we will focus on the mechanisms through which industry, EU policy interaction and the external environment affected the outcome.

Why has the EU Developed an 'EU Engineering' Renewable Energy Policy?

Industry: Triumph of the Economically Subordinate Renewable Energy Industry

When we turn to the role of industry, we find two striking features. First, hardly any industry actors promoted renewable energy in the pre-1997 period. The renewable energy industry did not yet exist, and the utilities did not show any interest in renewables. This means that the emergence of the EU renewable energy policy cannot be attributed to any influence from industry. We must search elsewhere for the mechanisms that made this specific EU policy possible in the first place.

Second, after 1997 we see a strong and increasing involvement on the part of industry, from the traditional electricity industry as well as from the renewable energy industry that had emerged during the 1990s. The renewables industry supported fairly strong European centralization of control, but was initially sceptical to market thinking. This scepticism increased over time. When the utilities started to show interest in renewables, they favoured the development of a common European market instrument. Hence, the two industries perceived their economic interests very differently: the electricity utilities preferred a Single European Market outcome, while the latter tended toward what we have called 'EU engineering'. The EU engineering policy outcome in 2010 is far closer to the position of the renewables industry than that of the electricity industry. In the following we explore whether this is the result of the one industry having had more influence than the other. Has the renewables industry simply been more successful than the electricity industry?

We begin with the 'instrumental governments' mechanism. Correlations between the positions of nationally dominant industries and government positions indicate that this mechanism has indeed been at work, but not always in the expected manner. Since industry actors lacked clear positions in the first period, this mechanism was obviously not in operation at that time. But by the late 1990s, Germany and Denmark had developed significant renewables industries and in many countries the traditional electricity utilities had started to engage in this policy area. Both industries stepped up their engagement after the turn of the millennium.

After 1997, we see a match between industry and national government positions in Spain, Sweden and to a certain extent Germany. Germany hosts two of the largest European utilities, both of which support a centralized market scheme, and has a significant renewables industry. Interestingly, Germany repeatedly sided with the latter and not the economically stronger utilities. Hence, we can conclude that the position of the German government reflects other concerns than the direct economic interests of the structurally stronger national industry. The government defended its traditional approach – the feed-in scheme – and not the viewpoints of the powerful utilities.

In Spain, the government shared the view of the dominant Spanish utility, Iberdrola. Interestingly, Iberdrola perceived its economic interests differently than did all other major utilities: even as a liberalized, Europeanized utility, it chose to defend the Spanish feed-in scheme. Also the Swedish government shared preferences with its leading national utility, Vattenfall. It seems as also Vattenfall defended the traditional national scheme, but since Sweden had introduced a green certificate scheme this led Vattenfall to support a pan-European market scheme. Germany, Spain, Sweden – in none of these cases do the government positions seem to reflect any one-way pressure from commercial actors. Instead, as we will discuss in relation to the New Institutional (NI) perspective, it seems that companies as well as the governments promoted the traditional national steering methods.

With respect to the two other member states in focus, UK and Poland, we see no close match between industry preferences and member- state positions. For a long time the UK was the prime promoter of market-based schemes. Since the UK lacks a strong national electricity industry, this cannot be interpreted as a result of national industry influence. Rather, the UK positions seemed to reflect British energy regulation traditions. Moreover, the UK changed its position in 2008, apparently without any clear backing from national industry interests. On the other hand, we may speculate that the absence of strong national utilities made it easier for the government to shift its position.

Initially, Poland had little interest in the development of EU policy on renewable energy, but it eventually took on an important role. As the Polish utilities do not seem to have expressed any clear views on this issue, the positions of Poland cannot be interpreted as resulting from national industry pressure. However, there was a link to Polish interests in another issue-area, the ETS – and this can therefore partly be interpreted as in line with Liberal Intergovernmentalist (LI) views on interaction, discussed later.

We can conclude that even though the economic interests of industries have been central in discussions on renewable energy policy, the EU member states have to a very limited extent acted as the instruments of their own dominant industries. Hence, it does not seem as if the LI perspective provides an apt mechanism for industry influence.

Turning to the NI perspective and the institutional feedback mechanism, over time we have seen that two distinct organizational fields – each centred on a specific industry – came to dominate policy development. The liberalized electricity utilities were central in the field of European electricity production, but the field also encompassed parts of the Commission, such as DG Competition, parts of DG Environment (later DG Climate) and parts of DG Energy. The governments of the

UK and Sweden, where energy markets had been liberalized and green certificate schemes had been developed, were affiliated with the field, but only Sweden had strong ties to a dominant European utility, Vattenfall.

Moreover, privatization and liberalization of the utilities made the ties between the utilities and the national governments weaker after the 1990s. Hence, it seems as if fewer governments took an active part in the field. Note also that the field lacked strong ties to MEPs. By the late 1990s the field was embedded in a market logic, and hence its actors promoted a Single European Market for renewable support. However, the field was not completely dominated by the market logic; and among the utilities, interest in renewable energy varied significantly. For instance, Iberdrola defended the Spanish feed-in scheme; and smaller utilities, like the Polish coal and electricity industry, showed little interest in renewable energy at all.

The renewable energy industry emerged from the feed-in support schemes of Germany and Denmark. This industry was embedded in a European field that can be seen as encompassing the German and Danish governmental organizations, and eventually also other prominent governments like that of Spain. Parts of DG Energy and the European Parliament had central positions in this field, and shared the field's focus on the logic of technology development. The field emerged in the 1990s, but became more unified and more coherent after the turn of the millennium. It seems as if the conflict over the first EU renewables directive was important for the development of the field: it consolidated the field by creating stronger ties between the actors, spurring more unified support for the logic of technology development and diffusion of the feed-in model. Most importantly, the two organizations EREC and EUFORES, both established while deliberations on the first renewable energy directive were underway, ensured coordination between Brussels-based and national actors. Since the late 1990s, the institutional logic of technology development was clearly dominant, and grew increasingly strong over time.

We may conclude that the organizational field of renewable energy encompassed more of the central EU level actors, was better coordinated internally and was characterized by stronger institutional unity than the field of electricity production. Moreover, the renewables field seems to have become more unified over time, whereas the electricity field became weaker. Because the two fields were characterized by different logics, they applied different criteria in assessing the efficiency of schemes. Actors in European electricity production preferred a scheme that favoured the most profitable projects in renewables, whereas the renewable energy actors preferred state funds to be directed towards the technologies most in need of support. Both fields were able to produce policy recipes in line with their dominant logic.

But the renewables field had one clear advantage: its favoured measure, feed-in schemes, had already been operating for years. In the late 1990s, the market idea of the utilities was merely a theoretical construct. This idea then went through a process of trial and error, but only Sweden developed a successful scheme. Moreover, the feed-in schemes diffused rapidly and had dedicated supporters in central European governments, whereas some of the core actors in the field of electricity production, like the British government, shifted their stance underway. This difference in institutional unity had a major impact on the EU deliberations. Note, however, that both fields supported a stronger centralization of control, and that this contributed to make member states without heavy stakes in developing renewable energy more positive to the introduction of binding national targets and greater Commission authority in this area.

Thus, institutional feedback mechanisms seem to have contributed to give the renewable energy industry a superior role in the EU policymaking processes. This can help to explain the 'technology-development' character of the policy outcome. First, internal institutional feedback effects within the field of renewables made it more united in supporting the logic of technology development. Second, institutional feedback effects in the field of electricity production made the field less coherent over time: energy market liberalization and privatization weakened the ties between utilities and governments, and the failure of many of the early market schemes reduced the credibility of this policy recipe. Institutional feedback provides a good explanation for why the renewable energy industry came to have more success in the EU policy development than the utilities, but it cannot fully explain how the renewables actors managed to garner support from actors outside the field, like Poland, the UK and other important member states. *Entrepreneurship* must be taken into account in order to understand this, as we shall see.

The Multi-level Governance (MLG) perspective leads us to focus on actors equipped with extraordinary intensity or skills for engaging in the policy processes. Do we find the network entrepreneurship mechanism at work? In the early years of the 2000s, renewable energy industry actors engaged in Brussels lobbying. At that time, the organizational field of renewables was not very strong. The German government was a powerful actor, and the renewables industry and activists seem to have used their German connections creatively in working to hinder the development of a market-based support scheme. Moreover, their networking resulted in the creation of more stable Brussels organizations: the EREC umbrella and EUFORES. This appears to have helped to prevent the first renewable energy directive from emerging with a stronger market character. In addition, it strengthened the European organizational field. Hence, the field of renewable energy had a stronger position during the second policy processes that led up to the 2009 Directive.

By that time there were far more entrepreneurial actors on the Brussels scene than in the first process: in addition to the German government and the renewables industry, also the Spanish government and prominent MEPs acted as entrepreneurs at this stage in time. Most important, MEP Turmes managed to be designated as rapporteur – an excellent foundation for entrepreneurial inputs. EUFORES and EREC helped to provide Turmes with capacity to perform extensive multi-level lobbying, reaching member states that had no prior affiliation with renewables.

In addition, the renewable energy actors were skilled, in the sense that they acted strategically, and deliberately chose to approach important governments that were initially not part of the renewable energy field. This may well help to explain why UK and Poland changed their positions and abandoned the market idea. Note also that the special climate package procedure gave the rapporteur extraordinary leeway to steer the process, and Turmes exploited this fully. Hence, we may conclude that the industry performed important network entrepreneurship throughout the first decade of the 2000s, but also other actors, and particularly MEPs, came to play a crucial role as time went on.

The entrepreneurial activities of the European electricity industry were not as successful. First, German utilities tried, unsuccessfully, to use the general regulations for state aid and the European Court of Justice as instruments to promote a market scheme. Second, they failed to create inclusive networks like those of the renewables actors, creating strong ties to member states and MEPs. Third, they were not able to sustain the Commission's internal support to the market idea. True, the market promoters in the Commission appeared as keen on promoting the issue, but the number of technology-development supporters in the Commission seems to have increased over time.

For one thing, the utilities appear simply to have been less active multi-level entrepreneurs than the renewables industry. But lack of intensity is not the full explanation: their entrepreneurial strategies failed, again and again. These repeated failures cannot be attributed to lack of entrepreneurial energy and skills. Rather, the nature of the organizational field of renewables created a better foundation for entrepreneurial achievements than did the field of electricity production. We must also recall that both camps sought to achieve stronger policy centralization, and here the market supporters in the Commission performed successful entrepreneurship: through dialogue with governments they managed to create a formula acceptable to the member states. All in all, we may say that network entrepreneurship enhanced the influence of the renewables industry on the policy outcome, even though it is difficult to distinguish clearly between the importance of entrepreneurship and institutional feedback mechanisms.

On this backdrop we can conclude that the renewable energy industry had more clout in policy development than the electricity industry. Renewables actors had superior entrepreneurial energy and skills during the deliberations on the first renewable energy directive adopted in 2001. They also performed successful entrepreneurship when the second renewable energy directive was developed, but their success in the first process also produced institutional feedback effects that enhanced their power basis this time around. Hence, thanks to both institutional feedback effects and superior entrepreneurial efforts, the renewable energy industry was able to have more impact on the EU policy outcome than the electricity industry.

Policy Interaction: More Marked by Failures than Successes

Renewable energy policy has a quite long European political history, but it was the member states, and not the EU, that took the lead in developing more specific measures and support schemes. The frontrunner countries Germany and

Denmark established feed-in schemes in the 1990s. In parallel, the EU introduced the ALTENER programme, focusing on the development of renewable energy technology and offering minor R&D funding. ALTENER was developed together with the SAVE directive and programme for energy efficiency, the latter having a similar technology-development profile. At this point, since efforts to introduce a carbon tax had failed, these two policies formed in reality the core of the EU's climate policies.

When work on a more elaborate EU renewables policy was initiated in the late 1990s, the development of such a policy was not linked to EU energy-efficiency discussions but more to the evolving discussion on the ETS. Grounded in the same market thinking that underpinned the ETS initiative, the Commission proposed to create a common, harmonized market policy for renewables. The 2001 renewables directive contained some weak market elements, but on the whole it had a 'local loading' character, giving the member states significant leeway to develop support schemes in accordance with their own preferences.

Then DG Environment increasingly realized that the ETS was not strong enough (in the form of producing a stable and sufficiently high CO_2 price) to create forceful economic incentives for the development of renewable energy, at least not in the short run. Hence, Commission officials with ETS responsibility once more initiated a market-based renewable energy policy, designed along the same lines as the ETS – but, as we have seen, they failed to introduce market elements in the final policy outcome. What does this mean when we look more specifically at interaction mechanisms?

We begin by searching for functional interaction, based on the Liberal Intergovernmentalist perspective. The initial link to EU energy efficiency does not appear to have influenced the distribution of power among the member states or policymakers in any way that affected the outcome on renewable energy: EU energy efficiency policy was weak and had few practical consequences. Later, it can be argued, the fact that the carbon price had been too low to spur development of renewable energy spilled into the debate on a policy for renewable energy, supporting the development of a stronger EU policy here.

However, it does not seem as if this shifted the distribution of power among member states or other policymakers to any great extent. For instance, Germany was a powerful actor both in the ETS and the renewable energy processes, but this was due to its position in both policy areas – as a major emitter (and hence with many allowances) and a major renewables producer. Further, initially the UK had been a supporter of the ETS, but the initiation of this system did not enhance Britain's ability to push for a market-based renewable energy scheme. Moreover, the UK shifted its stance in the latter case.

Do we see more bargained interaction? – not in the first policy development processes. The 2001 Directive was developed and decided largely on its own, and not in a package along with other climate-policy issues. This makes it unlikely that we will find much bargained interaction by digging deeper. With regard to the revision that led to the 2009 Directive, process insiders indicate that renewables were not part of any horse-trading in the final negotiation phase. The main renewables issues were in fact settled before the final bargains took place in the European Council.

It is, however, worth noting that Poland joined the 'compromise camp' in the summer of 2008, in practice blocking the possibility of introducing any forceful market elements in the revised renewables directive. Poland appears to have done this because it hoped that would strengthen its campaign for special treatment of and continued free allowances to its coal utilities in the ETS (and in this it had considerable success). Although it may be argued that Poland's tactics did influence the renewables outcome in 2008, we still maintain that the overall importance of bargained interaction was low.

Turning to institutional interaction, we see an initially striking similarity between the steering methods of EU policy on renewable energy and EU policy on energy efficiency, both underpinned by a technology-development logic – but we would hold that this similarity was not a result of interaction. A more reasonable interpretation is that the similarity came about because the actors that promoted these policies (primarily from DG Energy) were embedded in a logic of technology development.

From the late 1990s and onwards we see an increasing interface between the ETS and renewable energy policy processes, but with respect to actual policy outcomes, the market logic that underpins the ETS did not win much ground within renewables policy. Note that Commission officials working with the ETS tried to persuade member states as well as other actors that the policy on renewable energy should follow as similar institutional logic as the ETS. Although they devoted significant energy and creativity to such 'persuasion entrepreneurship', they did not succeed in justifying the market approach as a legitimate steering method. The market entrepreneurs did, however, also develop a key for target distribution, which they initially planned to combine with the introduction of a pan-European certificate scheme. Since the targets were accepted, they contributed to underpinning the centralization of this policy outcome, but failed in their main project: to persuade member states to opt for a market design.

Overall, the main impression is that the renewables process developed very much according to its internal logic. It was characterized by what can be seen as policy-internal path-dependency: the initial steering method of the German and Danish schemes spread among member states. The frontrunner member states and the renewables industry collaborated well and were able to block a shift towards a market approach. Due to the strong feedback effects from the emerging European field of renewable energy policy, market supporters were not able to get the market logic behind the ETS to spill over into the area of renewable energy. We do see some significant examples of functional interaction, but in the end this had little impact on the actual policy outcome. In fact, there were far more examples of failed than successful interaction efforts.

External Environment: Enabling the Commission to Perform Effective
Institutional Entrepreneurship

Important EU decisions of renewable energy policy have correlated with main
decisions in the global climate regime. As mentioned, this pattern was repeated
three times: in the early 1990s, at the entrance to the 2000s and then again in
2007–2009. Moreover, the EU has generally stepped up its policy development
before the crucial decisions are made in the global regime; afterwards, the policy
is specified in order to fulfil international obligations. Can any of the three
perspectives provide the mechanisms to explain this?

We begin with possible coercive pressure. The case of renewables provides
one example of a situation in which global market conditions played a key role:
the very beginning of European renewable energy policy was sparked by the
international oil crisis of the early 1970s. This was a precondition for the later
development of an EU policy on renewable energy, but the immediate EU response
was surprisingly weak, seen in relation to the magnitude of the oil shock. Instead,
it was the individual member states that reacted; and this did not affect the EU
policy process until many decades later. In addition, growing concerns for the
energy security of the EU represent a change in international market conditions,
but the EU could just as well have responded to this situation by stepping up its
production of fossil fuel or nuclear energy. The Kyoto commitment did require the
EU to develop climate policies, but the EU could have responded by developing
other policy measures. Hence, we do not find that coercive pressure had much
influence on the outcome.

Turning to possible institutional diffusion, it seems puzzling that the most
important EU action tends to happen before, rather than after, crucial global
summits. Moreover, neither the United Nations Framework Convention on Climate
Change (UNFCCC) nor the Kyoto Protocol focuses specifically on renewable
energy, so the global climate regime cannot explain why the EU should choose
to meet the climate challenge with renewable energy policy. Furthermore, cost-
effectiveness and market thinking dominate the climate regime, so that cannot
shed light on why the EU developed 'EU engineering' renewables policy. Hence,
it does not seem as if the EU policy outcome was particularly affected by the
mechanism of institutional diffusion.

Finally, does the correlation between the two spheres result from international
entrepreneurship? At each of the three peaks in EU policy development, EU-
internal actors, and most prominently the Commission, have argued that the swift
development of ambitious climate policy would make it possible for the EU to act
as a global climate leader. This relates to how Commission officials, and to some
extent member-state representatives, have linked the situation in international
negotiations to the EU climate discussion, as seen in Chapter 3.

In the early 1990s, the Commission argued that the EU should develop
substantial internal renewables policies in order to act as a global climate leader.
This contributed to the creation of the ALTENER programme and symbolic policy

statements, representing the starting point for the EU renewable energy policy development. Although the initial results were meagre, it did emplace renewable energy as a core component of EU climate strategy.

In the run-up to the Kyoto negotiations, the Commission repeated this argument, now with far greater vigour. At this stage, the environmental movement as well as the renewables industry actively supported this view, applying even clearer rhetoric. Interestingly, these actors were able to use Kyoto outcome as an incentive for the development of a stronger EU policy on renewables, even though the Kyoto Protocol scarcely dealt with renewables at all. This normative pressure may well have contributed to the development of the first EU renewable energy directive, but the Commission was not able to use this line of argument to achieve stronger centralization of control.

References to the global negotiations played an even more significant role in the run-up to the 2009 Copenhagen climate meeting. The big member states, like Germany, France and the UK, jousted for status as the most prominent global leader in the fight against climate change, and this gave the Commission room to argue that extraordinary measures were needed in order to develop a strong EU climate policy in which renewable energy played a vital role. As explained in Chapter 3, this resulted in the creation of an extraordinary rapid procedure for deliberation of the EU climate package. In turn, this procedure created a unique window of opportunity for the actors that supported a stronger EU policy on renewable energy. More specifically, it made it easier for the Commission and the renewables industry to create broad support for a more centralized policy. However, it did not help the market promoters in the Commission to gain support for a market-based design.

The entrepreneurship that was repeated three times had a clear institutional flair: the entrepreneurs used the global situation to spread the view that carbon mitigation was a pressing concern and to create the understanding that the internal policy actions of the EU were imperative to the global outcome. The failed Copenhagen meeting clearly illustrates that EU internal policy development was far from sufficient to 'save' the global process, but this could not be known in the period prior to the summit. Hence, this kind of entrepreneurship was not about creating networks and changing the information basis for specific decision-making situations, as highlighted by the MLG perspective. Rather, the entrepreneurs succeeded in affecting policymaking because they were able to influence how other EU policymakers understood specific situations and their role in these situations (see Finnemore and Sikkink 1998, Fligstein 2001, March and Olsen 1989).

On this backdrop we can conclude that developments within the global climate regime were coupled to the development of EU policy on renewable energy through a special kind of *international institutional entrepreneurship*, and it was due to this mechanism that the global backdrop contributed to create a steady path towards stronger centralization within this issue-area.

Conclusion: Crucial Combination of Institutional Feedback and International Institutional Entrepreneurship

This chapter has tracked and explained the emergence and development of an 'EU engineering' policy for renewable energy. Industry played no role in the emergence of this EU policy area in the pre-1997 period, but it had major impact on later policy development, leading to an EU engineering outcome. Policy interaction has not been very important: certain actors tried to persuade renewables policymakers to base the EU renewables policy on the ETS policy recipe – a Single European Market – but failed in their attempts to induce policy interaction. This 'persuasive interaction' did, however, contribute somewhat to the increased centralization. The external environment has been important, particularly by enabling entrepreneurs to legitimize a shift towards increased centralization of control in this issue-area.

Unlike what would be expected by the LI approach, the electricity producers (the economically dominant industry) do not stand out as important policy drivers: the member states did not act as instruments for the electricity producers. Rather, the gradually more unified and coherent organizational field of renewable energy created increasingly stronger institutional feedback mechanisms for a centralized EU policy characterized by technology development. In addition, the renewables industry, together with certain MEPs and some member-state representatives, exercised successful network entrepreneurship, underpinning centralization and a technology-development steering method.

Further, the global climate regime has not affected the development of renewable energy by coercive mechanisms: instead, it has created opportunities for entrepreneurship. Through reference to critical situations in the global climate regime, the Commission, with support from the renewables industry and the environmental movement, managed to make it appear crucial for the EU to develop increasingly stronger and more centralized policy on renewable energy. This also contributed to the initial emergence of this policy area. Due to the strong institutional flair, we have called this 'international institutional entrepreneurship': the entrepreneurs have used international developments to change other actors' perceptions of decision-making situations and to legitimize greater centralization.

We can conclude that it was the operation of institutional feedback, network entrepreneurship *and* international institutional entrepreneurship that enabled David – the relatively weaker renewable energy industry – to beat Goliath – the relatively stronger industry electricity producers – in the case of EU policy on renewable energy.

Chapter 6

Carbon Capture and Storage (CCS): *Carpe Diem* Entrepreneurship?

Introduction

From the early 1990s, the EU climate policy debate had centred on carbon pricing, energy efficiency and renewable energy. Then, after the turn of the millennium, Carbon Capture and Storage (CCS) entered the EU scene as a new climate solution. CCS has been described as 'a suite of technological processes that involve capturing carbon dioxide (CO_2) from the gases discarded by industry and transporting and injecting it into geological formations' (European Commission 2008e). In the course of only a few years, CCS changed from being an issue only for especially interested researchers and oil-industry engineers, to a central element of the 2008 climate package.

The EU's CCS policy consists of a directive for the storage of CO_2, complemented by financial mechanisms to support specific pilot projects. The EU offers direct financial support of CCS along lines not followed in other climate technologies: the arrangement is highly centralized and relies on technology development. The Storage Directive (European Parliament and Council 2009c) provides member states with significant leeway, but the financial mechanism gives the policy area a definite character of EU engineering. On this backdrop, we take up two main questions in this chapter: Why did this policy come into being? How did it get an EU engineering character?

Interestingly, the EU's CCS policy appears to have emerged without the kind of heavyweight support from prominent industries or member states or sustained Commission support evident in other areas of climate policy. CCS has certainly been among the most controversial elements of EU climate policy (see for instance *Euractiv* 2007b). Environmental organizations are split on the issue. Greenpeace, for example, brands CCS as a 'false hope' that draws attention and resources away from key policy areas such as renewables (Greenpeace 2008). Other environmental foundations, among them the British E3G and Norway's Bellona, see CCS as a promising technology with a considerable potential for reducing global CO_2 emissions rather quickly (see Bellona 2010). Prior to 2005, interest in CCS among EU member states was almost non-existent. Also the Commission seemed lukewarm initially, but made an about-turn around 2005 and began presenting CCS as a necessary component in a more elaborate strategy for climate-change mitigation. Eventually, a small CCS lobbying community emerged in Brussels, including MEPs and environmental organizations. CCS also attracted increasing attention on the global scene, within the climate regime as well as in other forums for political collaboration, like the G8. Did EU entrepreneurs then seize the opportunity created by international developments, and exert what we may call 'carpe diem entrepreneurship' to put a more comprehensive CCS policy in place?

In the following we will pay particular attention to how industry, policy interaction and the external environment have contributed to shape the CCS policy. Industry in general has not been actively opposed to CCS, but only the oil industry paid much attention before 2005, and never as a priority issue. Was the EU CCS policy firmly rooted in the oil industry, or did the policy emerge in spite of lack of interest from industry? Did industry actors play crucial roles in the CCS lobby community, or were other actors more important? In 2006 there were loose calls for specific economic incentives for CCS to be established, related to the ETS. Did subsequent policy interaction decisively shape the outcome of this process? On the global stage there were various CCS developments in the USA, and special CCS reports were produced under the auspices of the climate regime. Did such developments shape EU CCS policy?

In looking into these questions, we draw particularly on a Multi-level Governance (MLG) perspective, to shed light on how entrepreneurship affected the policy outcome. In addition, Liberal Intergovernmentalism (LI) and New Institutionalism (NI) perspectives can help to explain how factors both external and internal to the EU in conjunction fostered the CCS policy outcome.

The CCS Policy Outcome: Strong 'EU Engineering' Flavouring

The EU CCS policy has two main components: the directive on 'geological storage of carbon dioxide (hereafter 'the Storage Directive': Directive 2009/31/EC), which provides a legal framework that allows CO_2 to be safely stored deep underground; and financial mechanisms to stimulate the construction and operation of up to 12 demonstration plants by 2015. Overall, the policy has a fairly strong 'EU engineering' tinge, due not least to the significant competency delegated to EU organizations in relation to pilot-project funding and the clear technology-specific support – and, arguably, a general lack of market thinking.

We begin with the steering method of the policy. The Storage Directive specifies the responsibilities and obligations of governments and companies, with explicit and binding juridical clarification of important technical matters related to CO_2 storage. For instance, it states that the objective of geological storage is permanent containment, and that storage in water is prohibited. A highly specific requirement is the obligation for new power plants to be 'capture-ready' (ready to deploy CCS) by setting aside space for capture technology.

As to the financing of the above-mentioned CCS demonstration projects, a main element is the formal link established to the revised EU emissions trading system (ETS) post-2012: first, 300 million allowances from the New Entrants Reserve (NER) was set aside in 2008 to support up to 12 CCS demonstration projects and projects demonstrating renewable energy technologies: the NER 300 fund (see European Parliament and Council 2009b, Art.10a.8. See also Chapter 4). It has subsequently been clarified that the NER 300 is intended to contribute around half of the means necessary to fund eight CCS projects and 34 renewables projects (European Commission 2010c).

Because CCS demonstration projects are far more costly than projects in renewables, the financial mechanisms must be regarded as an important part of the EU's CCS policy. Each country will get minimum one and maximum three projects. By the February 2011 deadline, applications for 22 CCS projects (and 131 renewables projects) had been submitted to the Commission (*Point Carbon* 2011b). Moreover, the Commission decision on NER 300 specifies detailed technological criteria for the selection of CCS and renewable projects (European Commission 2010c, Art. 8). Applications and auctioning of allowances will be handled by the European Investment Bank, interacting with the Commission and member states (NER 300 2012).

Project selection is based mainly on specific technological qualities, with economic criteria a more secondary concern. For instance, the Commission is not required to choose the least costly projects, or projects based on technology solutions most likely to become profitable in the immediate future. The size of CCS funding depends on price developments in the ETS market, but the CCS policy as such is not characterized by market thinking or logic. CCS funding relies more on technological than on economic criteria. As the link to the ETS and this financial support did not become fully established before 2011/2012, there was a need for some short-term financial incentives as well. In the spring of 2009, the EU assembled a financial stimulus package for 2009 and 2010 to counter the effects of the global financial crisis. Of the four billion euro in the European Energy Programme for Recovery, around one billion was earmarked for CCS demonstration projects in six countries: Germany, the UK, the Netherlands, Spain, Italy and France (*European Energy Review* 2011).[1]

Do we find centralization of control in this policy area? The 2009 Storage Directive merely guides member-state practices with respect to storage permits, allowing substantial leeway for national adjustments. Member states are to determine the areas to be made available for storage and the conditions for site use. With regard to storage permits, the Commission is given a watchdog role, reviewing draft permit decisions and providing opinions that the competent authority will need to take into account (Directive 2009/31/EC, Art. 10). Storage permit regulations are a precondition for national CCS activities, and the Directive promotes CCS as a technological climate solution, without obliging the member states to realize or support CCS projects. However, the related EU financial measures more clearly promote CCS and provide the EU organizations with significant authority.

---

1   By 2011, six projects had been granted funds: five have received 180 million euros, while one has received 100 million. See *European Energy Review* 2011.

We conclude that, due primarily to the financial mechanisms like the NER 300 fund, the EU's CCS policy smacks strongly of EU engineering. First, it promotes technology development and relies on technology criteria, not economic criteria or market measures. As to the distribution of competence, the financial dimension of this policy involves significant central steering. Since the member states are expected to contribute half of the expenditures and oversee the actual construction of pilot plants and the implementation of EU CCS policy, there are some clear limits to EU central control here. The storage part of EU CCS policy has more of a 'local loading' character. Yet, the CCS policy has a more heavily centralized distribution of competence than many other areas of climate policy. Taking particularly the financial mechanisms into account, we would nonetheless hold that the policy as such shows a clear imprint of EU engineering.

The CCS Policy Story: From Nerdy Issue to Name of the Game?

1997–2004: CCS Emerges as Possible Solution

CCS has a fairly brief political history within the EU, although the US oil industry had used CO_2 (as an alternative to water and gas) for enhanced oil recovery since the 1970s (ZEP 2006: 13). In contrast to carbon taxation and pricing, or renewable energy and energy efficiency, CCS was not debated as a possible solution in the first wave of EU climate policy initiatives in the early 1990s. Neither was CCS mentioned in the EU's first, more elaborate response to the Kyoto Protocol, the 1998 Commission Communication on Climate Policy (European Commission 1998b).

In 2000, the first European Climate Change Programme (ECCP I) mentioned CCS as one of several possible 'common and co-ordinated policies and measures on climate change' related to energy supply (European Commission 2000a: 11). The ECCP working group on energy supply discussed CCS, but ended up not proposing any EU policy measures. CCS was described as a somewhat immature 'post-2010' technology (European Commission 2001a: 48). Nor was CCS mentioned as a possible abatement option in the 2003 Emissions Trading Directive (European Parliament and Council 2003). Our interviewees note that it was very hard to get EU officials to pay attention to CCS in this period, particularly DG Environment. For instance, a civil society representative mentioned Commission officials who claimed that CCS was an unrealistic and somewhat 'crazy' idea.

In the early 2000s, the R&D units of European oil corporations – particularly British BP, Dutch Shell and Norwegian Statoil – began exploring CCS possibilities (see BP 2009, Boasson 2005, ENI 2009, Total 2009). French Total and Italian ENI engaged more moderately. Due to the relatively flat structure of the European oil majors, the significant size of their technological units, and their prior experience with enhanced oil recovery, engineers from the oil industry were able to develop the technological foundations for CCS. Within a few years, the European oil corporations had recognized climate change as the most pressing social issue confronting the industry, but corporate headquarters promoted emissions trading as the key policy solution (Boasson 2009, Boasson, Bohn and Wettestad 2009). CCS was an issue mainly for the R&D departments.

At the time, CCS was discussed chiefly in relation to petroleum exploration and processing, not as a way of mitigating emissions from stationary energy production. This can be seen as a symptom of lack of interest on the part of the big European power-producing utilities. Sweden's Vattenfall was the exception, initiating CCS research already in 1996 and launching a CCS pilot plant in Germany in 2005 (Buhr and Hansson 2011). Our interviews indicate that despite the European utilities' heavy dependence on coal, no other utilities engaged in the CCS issue at the time. Among the environmental organizations, Greenpeace opposed CCS and, with the exception of the Norwegian organization Bellona, no environmental organizations actively supported the development of an EU policy on CCS.

Overall, the governments of EU member states perceived CCS as a R&D issue, not one of policy. Germany, the Netherlands and the UK initiated CCS research initiatives, but there was little related policy development (Meadowcroft and Langhelle 2009a). Germany established two CCS-related research programmes in 2000 and 2003, but even though almost half its electricity was generated by domestic coal, CCS remained a non-issue in the country's climate policy (Praetorius and von Stechow 2009: 139, 146–47). The Netherlands launched a CCS research programme in 2004, but otherwise the issue attracted little interest (Vergragt 2009). In the UK, CCS was discussed as a possible 'clean coal solution', and actors in the British oil industry voiced a certain interest in CCS linked to enhanced oil recovery, but political interest in CCS was meagre (DETR 2000, Scrase and Watson 2009).

In other EU member states CCS was basically a non-issue. For instance, studies of Spanish and Polish climate policy indicate that it was simply not on the agenda (Costa 2011, Jankowska 2011). In Sweden, some actors voiced criticism of Vattenfall's CCS engagement, claiming that it prolonged the use of fossil-fuelled power production, but otherwise CCS received scant attention also there (Buhr and Hansson 2011).

Looking beyond the EU, interest in CCS as a policy measure within the global climate regime was still quite moderate (see Meadowcroft and Langhelle 2009b). The Kyoto Protocol set the development of market measures and hence the promotion of least-costly mitigation measures as a main theme – not the development of new, large-scale technological solutions like CCS. However, other global developments provided certain incentives for development of CCS technology. In 2001, the Conference of the Parties to the UN Climate Convention requested the Convention's Intergovernmental Panel on Climate Change (IPPC) to assess 'the scientific, technological and socio-economic aspects of CCS' (Coninck and Bäckstrand 2011: 371). Also the International Energy Agency became involved, hosting CCS conferences and issuing several CCS reports in the early 2000s.

The USA now emerged as a firm proponent of CCS (Helm 2009, Stephens 2009, Coninck and Bäckstrand 2011). Following the US withdrawal from the Kyoto

Protocol in 2001, the Bush administration launched a new project: FutureGen. This aimed at constructing the world's first large-scale zero-emission coal-fired power plant, demonstrating CCS, hydrogen production and other advanced coal technologies at the same time. However, as one EU official interviewee notes, 'CCS had a credibility problem (…) The Bush initiatives were more negative than positive for the EU process because the environmental camp was against all that Bush was for.'

Another relevant EU-external development took place in a different issue-area – within the regulation of marine pollution and particularly the OSPAR Convention, which covers the Northeast Atlantic (see Wettestad 1999, Skjærseth 2000). This cooperation dates back to the 1970s, with the initial Oslo and Paris Conventions which became the OSPAR Convention in 1992. Our Norwegian interviewees note that it was Greenpeace's concern about a Norwegian initiative to conduct a limited storage experiment in the water column that caused storage of CO_2 to be placed on the OSPAR agenda in 2002. A clarification process was initiated and a workshop held in 2004. Further clarification work was initiated within an OSPAR Committee (Sæverud and Moe 2005).

All in all, we do not find clear and vocal proponents for the development of an EU CCS policy in the period 1997–2004, within the EU or in the external environment. True, technological researchers, particularly within the oil industry, were involved in developing CCS technology, but they did not call for EU policy. The Commission had begun to develop a slight interest in CCS as a possible future climate measure, but this was on the fringes of the broader climate policy debate. It was after 2005 that the radical changes occurred.

2005–2010: Climate Hype Underpins CCS Entrepreneurial Campaign

CCS now began to gain substantial momentum as a political issue. The year 2005 marks a shift in the attitude of the Commission. This was the year when international negotiations over a new international agreement for the period after the Kyoto Protocol started (Andresen and Boasson 2012). According to one EU official interviewee, 'I believe the international negotiations were important for the change in the Commission approach to CCS: this was the only way to deal with the challenge. Many [Commission officials] came out as CCS supporters at the time.' Another external factor working in the same direction was the IPCC special report on CCS, published in September 2005 (Metz et al. 2005, Meadowcroft and Langhelle 2009b: 6). The UN engagement and the scientific backing made it easier for climate-policy officials within the DG Environment to frame CCS as a credible measure. This was particularly useful since, as noted, Bush's CCS advocacy had given CCS a somewhat bad name in the EU.

The Commission's internal shift in attitude was also related to other developments, especially the greater attention to energy security. According to one EU official, 'particularly the unit that worked with clean coal within DG Energy was interested'. As mentioned in Chapter 3, the heightened focus on energy security was closely related to the EU's energy policy relationship with Russia. Yet, DG Environment was internally split on CCS, and our interviews with EU officials indicate that this split lingered on: 'there was a majority backing CCS in the Commission overall before there really was a positive majority within DG Environment', according to one interviewee. Climate-change officials were generally positive, but officials dealing with resources and water were more negative, because of the potential problems with storage and leakage, among other things. Eventually, CCS supporters were able to create a consensus on the importance of having strict control of the storage processes, including the development of procedures for use in case of accidents.

So it appears that there were some key Commission officials who were beginning to see CCS as a policy ingredient that could promote EU climate ambitions while also improving the EU's global economic and technological competitive position as well as regional energy security. In February 2005, the Commission announced a second European Climate Change Programme (ECCP II), noting that particular attention would be given to CCS (European Commission 2005c: 11). A CCS working group was established. It produced a report that emphasized 'the *urgent need* for the development of a policy and regulatory framework for CCS', and recommended that the Commission should outline a proposal for an EU CCS regulatory framework (European Commission 2006e: 7; our emphasis added). In April 2005, the DGs for Energy, Environment and Research jointly arranged a conference, 'Towards zero emission power plants'. At this conference, EU energy commissioner Andris Piebalgs told the audience that Europe should take the lead in developing the CCS technology (*ENDS* 2005b).

DG Research initiated the European Technology Platform (ETP) for Zero Emission Fossil Fuel Power Plants (ZEP) (Claes and Frisvold 2009). This included various branches of industry (utilities, oil and gas, equipment suppliers), science and research communities, and ENGOs.[2] The ZEP mandate included the drafting of a mission statement and two comprehensive policy documents – the Strategic Research Agenda and the Strategic Deployment Document. The ETP–ZEP established an Advisory Council and several working groups, involving 150 to 200 experts. The Commission's 2006 Green Paper on Energy further bolstered the Commission's new view that CCS could contribute to achieving both energy and climate security (European Commission 2006c).

Besides the shift in the Commission's attitude towards CCS, what changes were there within industry, the EU member states and the European Parliament? Concerning industry, the creation of ZEP led the R&D actors in the oil industry to focus more explicitly on the policy side of CCS development. The Norwegian ENGO Bellona had been given the role as secretariat of the ZEP, and used this position to encourage ZEP to launch proposals for EU policy. In 2006, ZEP issued its Strategic Deployment Document that recommended, among other things, the

---

2 ZEP industry members are companies, not industry organizations or other umbrella organizations.

creation of EU regulations on CO_2 storage and the creation of 'an early mover funding mechanism to support the development of 10–12 large-scale CCS projects which demonstrate a diverse range of infrastructure, technologies, fuels and storage locations' (ZEP 2006: 2). The document did not specify whether funding mechanisms should be created by the EU or by its individual member states.

The UK now took a more proactive position. Political consensus had emerged concerning national governmental funding of CCS. Interviewed EU officials indicate that, due largely to the greater international attention accorded CCS, the UK started to promote the development of a EU policy. As one interviewee notes, 'the UK wanted CCS, but this was much related to the international climate negotiations. The focus was very much on CCS in China' (see also Scrase and Watson 2009: 170). We do not find a similar rise in CCS interest in other member states, although the issue of onshore storage had spurred some controversy in Germany. In contrast to the UK, Germany hardly has offshore storage options (*Spiegel* 2011). The Parliament actively pressed the Commission to launch renewable energy and energy-efficiency initiatives (see Chapters 5 and 7), but none of these initiatives related to CCS.

As described in Chapter 3, the key new EU climate targets were adopted in March 2007. The Communiqué from the European Council included two main CCS elements and was a clear indication that member states had increasingly begun getting on board 'the CCS train' initiated by the Commission. First, the Commission was asked to establish a legal framework for CCS, to allow safe underground storage of CO_2. Second, and particularly important, it initiated a policy to 'stimulate construction and operation by 2015 of up to 12 demonstration plants'. Concerning the references to the demonstration plants, the wording is strikingly similar to the ZEP recommendations, but our interviewees disagree on whether it was the UK or the Commission that initiated this part of the Council decision. In any case, we note that the Council decision did not mention the development of any specific funding mechanisms to ensure realization of the demonstration plants.

Immediately after the Council meeting, the Commission started developing a CCS storage directive, in close collaboration with the ZEP platform. EU officials point out that ZEP was important for the technological aspects of the directive: 'it was very useful as a one-stop shop that was giving advice'. Interviewees, EU officials as well as civil society representatives, describe ZEP as a quite independent group. Brussels interviewees agree that, out of the nearly 40 Commission-initiated technology platforms, ZEP took on an extraordinarily proactive role in the development of EU climate policy. The oil corporations devoted substantial and increasing manpower to ZEP.

Moreover, the above-mentioned CO_2 storage clarification process in OSPAR led to the adoption of a set of regulations in June 2007. A set of principles with respect to regulation were produced – for instance, a prohibition against placing CO_2 streams in the water column or on the seabed: such streams would have to be stored in geological formations underneath the seabed (OSPAR 2007a, b).

Interviewed EU officials note that this legal and institutional clarification provided important inputs to the development of the CCS Directive. Furthermore, our interviewees agree that neither the Parliament nor the EU member states contributed much in shaping the character of the Storage Directive. With respect to the financial mechanisms, however, the situation was radically different: it was here that political action began to emerge after 2007.

The European Council decision in March 2007 had given the Commission a heavy workload, not the least with the development of a draft for a storage directive. DG Environment was *chef de file* of the CCS dossier, whereas DG Research had initiated the ZEP platform. DG Environment had little time to follow up the decision on stimulating the construction of pilot plants. According to one civil society interviewee, the Commission phoned the member states 'and asked them what they wanted to do with the CCS pilot project statement in the 2007 decision; the member states said they would have to think a little bit more about that, but no one phoned back.'

As a result, the 2008 climate package proposal from the Commission included few incentives for member states to promote CCS construction, besides proposing that power plants should be 'CCS ready' where technically and economically feasible. The ETS directive proposal noted that 'there would be no need to surrender allowances for emissions stored' (European Commission 2008g: 5). An accompanying Commission Communication discussed obstacles to establishing up to 12 demonstration plants by 2015 (European Commission 2008b).

It was increasingly realized that the ETS would not offer a carbon price high enough to underpin and justify CCS, at least in the short term. Earlier ECCP documents had described CCS as a technology that could be stimulated by the ETS, and had proposed including CCS as an eligible activity under the system (European Commission 2006e: 9). Furthermore, it had been indicated that successful CCS development would require the creation of an additional and specific support mechanism (European Commission 2007f: 6). Now in the ETS directive draft, the Commission aired the general idea of setting aside ETS allowances for climate mitigation purposes, like CCS – but this was only a loose idea (cf. European Commission 2008g, Art.10.3c).

At this stage, the UK-based environmental investment management firm Climate Change Capital came up with the idea of setting aside a certain number of allowances from the New Entrants Reserve in the ETS, and using the auctioning revenues to finance CCS pilot projects (see Hampton 2008). The idea attracted attention and was soon taken up by organizations E3G and Bellona. It also gained hold in British/Dutch Shell, the French power plant equipment producer Alstom, and Swedish Vattenfall (see Center for Public Integrity 2009). A CCS financial mechanism lobby network emerged, made up of environmental organizations, industry actors, MEPs, Commission officials and member-state representatives. Some of these actors also promoted the introduction of an emissions standard that would make CCS mandatory on all new CO_2 point sources of a certain size. As noted by a central renewable energy actor in Brussels: 'the CCS community was really fascinating. They grew out of nowhere and suddenly they were there! We, the renewable actors, have a lot to learn from their straightforward professionalism.'

Whereas all participants in the loose network promoted the idea of an EU financial mechanism, the environmental organizations and some MEPs also advocated the introduction of an emission performance standard that would make CCS mandatory for future fossil-fuel plants. Eventually the core actors gathered under the name 'CCS leadership coalition' (Bellona 2008). As noted, the environmental movement had been split on the CCS issue. In 2008, Greenpeace published a highly critical report (Greenpeace 2008). However, WWF gradually became more positive, joining E3G and Bellona (*Euractiv* 2008b). Many core lobby actors held positions within ZEP, but they operated independently of the technology platform. The growing number of participants in ZEP made it harder to develop common positions. Most importantly, power producers stepped up their involvement in ZEP, and interviewees point out that there was a split between the utilities and the oil corporations. ZEP thus played a less active role at this stage.

Similar to the oil companies, the utilities were embedded in a financial market logic, and thus had a natural focus on the ETS. However, the R&D units had less prominent positions in the power-producing industry, and the utilities had not developed a tradition of environmental legitimization strategies and responses like those of the European oil corporations (see Boasson 2011). Hence, the utilities were far more reluctant when it came to promoting a new climate mitigation technology, and were particularly sceptical to the introduction of an emission performance standard. The oil industry was becoming more and more interested in CCS, also at corporate headquarters level, especially due to the related possibilities of enhanced oil recovery. But the utilities remained lukewarm. Vattenfall assumed a far more proactive role than the other utilities, standing out as a central CCS lobbyist in 2008 (Buhr and Hansson 2011).

Also the mood in the Parliament began to shift in a more CCS-friendly direction from 2007 on. This change was related to the increased global focus on CCS as a crucial element in future low-carbon energy systems, with specific relevance for India and China (see also Coninck and Bäckstrand 2011). Still, mixed feelings as to CCS lingered on, with the Greens as the most sceptical group. Chris Davies, from the Liberal group, was appointed Parliament Environment Committee Rapporteur for the CSS Directive in February 2008. Davies summed up his view on CCS in these words: 'I hate CCS...It is just that I hate coal more. We have to promote CCS. China, India and the US need to realize that they will have to pay a lot more if they want to use coal' (Friends of Europe 2008: 24).

Davies warmed swiftly to both the idea of a CCS financing mechanism and the proposal for a CCS emission standard. In the first debate in the Parliament conducted in early May 2008 he put forward two main proposals. First, he proposed that up to 700 million allowances be set aside in the ETS post-2012, to kick-start CCS. Second, all new fossil-fuelled power plants should be 'CCS ready': no carbon-emitting power plant should be given green light from January

2015 unless 90 per cent of its CO_2 emissions could be captured and stored. Davies received both support and criticism for these proposals (*ENDS* 2008e).

In early June, the Parliament Rapporteur for the ETS revision, Avril Doyle, followed up on CCS Rapporteur Chris Davies' call for setting aside allowances within the ETS New Entrants Reserve for funding CCS. This spurred a dynamic within the Parliament, and in early July the ETS and CCS Rapporteurs, together with Linda McAvan (also from the Environment Committee in the Parliament), proposed that up to 500 million allowances should be set aside within the ETS (*Point Carbon* 2008b). From this point, the Parliament stood forward as a network hub for actors seeking a stronger link between the ETS and CCS policies. In the same month, CCS received renewed global attention when the G8 countries stated that they 'strongly supported' the launching of 20 global large-scale CCS projects by 2010 (G8 2008). A Parliament Environment Committee meeting in September endorsed the Davies–Doyle–McAvan proposal. At this meeting, Davies also put forward the idea that, after January 2015, CCS should be mandatory for power plants over 300 MW that emitted more than 350 g CO_2 per kWh (*ENDS* 2008g).

The Commission was divided in its views on the character of the link between the ETS and the CCS policy, with DG Environment more reluctant than DG Energy (*Euractiv* 2008a, *ENDS Report* 2008). This was confirmed in our interviews: Commission officials were reluctant particularly because an additional funding mechanism and an emission performance standard 'would not work well with the ETS'. Hence, it was primarily the Parliament that championed greater EU steering and central control here, not the Commission. Interviewees emphasize the importance of the close cross-party collaboration between the two rapporteurs. According to one central actor, 'nothing would have happened if the Parliament had not acted on the issue. Chris Davies was crucial, but Avril Doyle was maybe even more important than Chris – she had the money!'

It was a deliberate strategy to use the emission standard idea to leverage a deal. As one civil society interviewee explained, 'the emission performance standard was very important in order to get green votes'. In addition, for tactical reasons, CCS promoters included renewable energy early on in their financing proposal: 'we needed it in order to get it through the Parliament (...) Member states not so interested in CCS could get renewables instead.' Furthermore, one interviewed lobbyist noted: 'of course we changed the argument slightly depending on who we talked to. When we spoke to the centre-right it was all about the survival of the coal industry.' Hence, the argument that a strong EU CCS policy was in the interest of the coal industry was used by CCS promoters at an early stage.

Interestingly, this industry itself started to portray its own interests in this way only *after* the EU CCS policy had been adopted (see EURELECTRIC 2011). Even though this argument enjoyed some clout among conservatives in the Parliament, there was considerable reluctance also within this camp. However, reluctance among the conservatives in the Environment Committee was not as strong as in the rest of the Parliament. According to one civil society interviewee, 'because it was only the Environment Committee that voted, it was enough to get half of the

MEPs in this Committee on board. Avril Doyle knew that. Her ETS inputs would never have passed the plenary if it had been an ordinary process.'

So on 7 October 2008, the Parliament's Environment Committee adopted two important CCS reports. First, MEPs endorsed an amendment setting aside up to 500 million allowances from the ETS New Entrants Reserve for funding CCS demonstration plants. Second, Rapporteur Davies got almost unanimous support for an amendment proposing CO_2 output limits, that is, emission performance standards (EPS) on power stations built after 2015 (*Euractiv* 2008c, *International Environment Reporter* 2008). Both EU officials and civil servants emphasize that the voting in the Environmental Committee was a nerve-racking occasion. First, CCS supporters lost the financial mechanism vote by one vote, apparently because one person had left the room and the Polish MEPs were undecided. However, as one interviewee explained, 'according to the voting rules, the original proposal will come up for a vote again if the compromise falls (…) so we did what we could in order to change the voting the second time around'. Most important, Polish representatives were given promises that Poland would not get hurt – and then the majority in the Committee shifted their stance.

How did the member states react to these new developments? We can note a certain increased interest in CCS, particularly in countries that had established CCS research programmes in the first period: Germany, the Netherlands and the UK. Interviews indicate that this change was related to the global situation and perspectives, with arguments about the urgent need for clean coal in India and China. Initially, most member states remained silent, without any strong positions on CCS. As put by one interviewee: 'CCS was really such a small thing; few actors paid much attention to it. Many of the other issues in the climate package required much more attention.' Greece was the most vocal opponent, arguing that earthquakes could destroy storage facilities.

The Council of Ministers reached a first compromise agreement in late October. In this agreement, member states gave support to the Commission's proposal to make power plants 'CCS ready' where technically and economically feasible, and rejected the Parliament's call for specific emission performance standards. Furthermore, a majority opposed the Parliament's proposal of setting aside 500 million ETS allowances for funding CCS demonstration projects (*ENDS* 2008j). The stage was now set for EU trialogue negotiations, which commenced in early November (see Chapter 3).

Parliament Rapporteur Chris Davies toured the capitals of the EU member states promoting the Parliament's amendments. His efforts were coordinated with those of other CCS promoters who also visited major EU member states. Moreover, interviewees point out that the French electricity equipment producer Alstom lobbied the French Presidency in order to ensure that the Parliament's proposals 'remained on the table' during the high-level negotiations on the climate package. Of key importance in this context, the director of DG Environment, Mogens Peter Carl, joined the French Presidency (*Financial Times* 2008b). He had headed the development of the climate package, and several of our interviewees have emphasized that he was central in ensuring that the Parliament proposals stayed alive.

Member-state interviewees as well as EU officials note that the UK was the most outspoken proponent of an EU CCS policy that would include a strong financial mechanism. In 2008, this was the main negotiating point of the UK government in deliberations on the climate package. At this stage, the German government was split over CCS. Chancellor Merkel was definitely pro, whereas the environment minister was sceptical. Still, a joint statement from the German ministries of the environment, economics and research in September 2007 had characterized CCS deployment as 'necessary and possible' (Praetorius and von Stechow 2009: 147). But Germany was reluctant to a centralized EU CCS policy, preferring that projects be funded nationally and not by the EU (*Point Carbon* 2008c).

At first, Spain was not very interested in CCS, but the link to renewable energy brought into the funding mechanism made the government more supportive. Initially, Eastern European countries like Poland were somewhat sceptical, especially because they did not want money to be taken from the ETS New Entrants Reserve. According to interviewees, 'they felt that it was their money'. Eventually Poland became more positive. This change seems to have resulted from targeted lobbying to persuade Poland that it would be able to get CCS and renewable funding from the financial mechanism.

By mid-November 2008, a significant number of member states had changed their positions, and a majority in the Council supported some kind of CCS earmarking of ETS allowances. The Commission, and DG Environment in particular, argued for a 'sizeable reduction' in the number of earmarked allowances (*ENDS* 2008k). The French Presidency basically sided with DG Environment and proposed a lower figure, between 100 and 200 million allowances (*Point Carbon* 2008d, Reuters Planetark 2008b). Also the Parliament sought a compromise; Avril Doyle opened up for accepting 350 million earmarked allowances, instead of the 500 million requested earlier (*Point Carbon* 2008e).

EU officials and civil society interviewees give lively descriptions of dramatic last-minute negotiations with respect to the policy outcome. These individuals were not present in the room, but many had text message contact with the negotiating parties. Funding of CCS was one of the last unresolved issues. The discussions had begun with a proposal for 200 million allowances to CCS funding. No agreement seemed to be in sight, but the parties knew they would have to find a solution before the announced press conferences. Observers maintain that 'at the very last minute, the British dug their heels in and obtained an increase in the number of allowances from 200 to 300' (*Euractiv* 2008d). So the final figure ended up being 300 million allowances.

Like the other issues in the climate package, the CCS policy was developed in a special process, with only one parliamentary reading and the European Council as the key decision-maker, not the Environmental Council and the Parliament. This process seems to have been of specific importance in this case. As one EU official reflects: 'I am sure it would have been shot down if the case had followed the formal procedure (…) this would never have been possible if there had been two readings (…) When it comes to the urgency as such, it certainly did help. I do not think a compromise like that would have been possible in 2010.'

Main developments after 2009 include the elaboration of rules for the specific ETS funds for CCS and renewables, the NER 300. As laid out earlier in this chapter, these developments specified the outcomes, without changing the fundamental character of the results. Hence we will not go into further detail here. What we can note is that CCS lobbyists played important entrepreneurial roles in policy development, and that EU-internal and EU-external factors underpinned their entrepreneurial achievements. Let us now turn to the main mechanisms at work.

Why has the EU Developed an EU Engineering CCS Policy?

Industry: Joining the Bandwagon instead of Pushing It

We begin by asking how this policy area has been shaped by issue-internal dynamics within the EU, with a specific view to the role played by industry. How have the main industries positioned themselves? CCS had equally high relevance for the oil and the electricity industries, as it could contribute to cut industry emissions under a stricter climate policy scenario. The oil industry, with its storage expertise, became interested in CCS technology development already early on, although CCS was not central in the oil corporations' climate-policy inputs. Both the oil and the electricity industry may deploy CCS in order to cut their emissions, but the latter group began to show interest only after the major EU decisions had been made. Electricity equipment producers provide CO_2-capture technology, and one such actor – Alstom – played a certain role in the process of policy development.

The ZEP technology platform, initiated by the Commission, became an important meeting place for industries, MEPs and various environmental organizations. Eventually also an additional CCS lobby network emerged in Brussels. Arguments along the lines that 'CCS is crucial for the survival of the European coal industry' figured high in the final stages of policy development, but our research indicates it was primarily the environmental organizations and MEPs that argued in this manner, not actors from the electricity industry. Moreover, oil industry representatives generally played a far more proactive role in policy development than did representatives of other industries. The EU engineering outcome is also arguably closer to the preferences of certain actors within the oil industry – the research units – than to those of most other industrial actors. How then can these various perspectives and mechanisms help us to understand the role of industry in the case of CCS?

Starting with the relevance of the Liberal Intergovernmentalism perspective and the instrumental governments mechanism, we will first look for similarities in the positions of industry and member states, since such correlations would indicate that this mechanism has been at work. Between 1997 and 2005, three member states (Germany, the Netherlands and the UK) initiated national CCS R&D activities, but without encouraging the EU to develop CCS policy measures. The Netherlands and the UK are home to large oil corporations (Shell and BP): this probably influenced the development of public CCS research programmes, but we lack detailed empirical information on this. Interestingly, also Germany developed a CCS R&D policy – although it had no national industries involved in CCS. However, the three national governments focused solely on national R&D measures; it was the Commission that initiated the development of EU CCS policy in this period. Hence, any correlations between industry and member state positions in UK and the Netherlands seem irrelevant for explaining the initiation of a EU CCS policy, and we conclude that the instrumental governments mechanism was not at work in the first period.

From 2005 onwards, more industrial actors and member states developed EU policy positions on CCS. The Netherlands and the UK, hosting the most active CCS-promoting oil corporations (BP and Shell), most clearly supported a strong EU CCS policy. BP and Shell became involved earlier than the governments, so the positions of the two governments may to a significant degree have been affected by the industry. On the other hand, it seems as if the oil industry voiced their inputs to the CCS directive mainly through the ZEP platform, not through member-state governments.

Further, we note that Vattenfall, the only utility that promoted an active EU CCS policy early on, was unable to get the government of its host country Sweden to champion CCS actively. France played a key role primarily by using its EU Presidency position in the autumn of 2008 to keep the CCS financing issue on the table. True, French equipment producer Alstom lobbied for France to do this, but it was not necessarily the simple result of national industry pressure. It may equally well be interpreted as the result of entrepreneurial engagement from outside and within the Presidency secretariat.

Other member states with key roles in the policy negotiations lacked national industry backing for their positions. Germany, for instance, had a crucial role in the final phase of CCS policy negotiations, bolstering the EU engineering imprint. But Germany's position was hardly a simple reflection of national industry interests. Poland eventually emerged as a supporter of an EU engineering policy – and this appears to have been crucial for the final outcome. However, this most probably reflected external lobbying from various Brussels- and UK-based CCS policy promoters, not pressure from Polish industry. Thus we see that the instrumental governments mechanism did not set in motion the EU CCS drive. The mechanism might have been operating later on, particularly for the Netherlands and the UK. But we must take other mechanisms into account to understand the outcome more fully.

Turning to New Institutionalism, in order to examine the importance of institutional feedback, we will first examine the character of the organizational field(s) involved in policymaking. CCS policy affects primarily the oil and electricity industries. These industries can be seen as parts of two distinct organizational fields, since the dominant firms within the two industries differ in their national rooting, their affiliations with Brussels actors, and their traditions in climate policy involvement.

As seen in Chapters 4 and 5, European electricity production as an organizational field emerged during the 1990 and early 2000s and was increasingly characterized by market thinking and pan-European approaches. However, this chapter has shown that none of the actors in this field (be it EURELECTRIC, the large utilities, or parts of the Commission) engaged in the CCS policy discussion to any extent. The organizational field of petroleum production is dominated by the large European oil corporations, and thus bears a heavy imprint of market thinking. Neither of these corporations engaged much in the CCS discussions, but engineers from their R&D divisions took an interest in this abatement option.

Due to the significant size of the European oil majors' R&D units and their experience with enhanced oil recovery, an increasing number of technical experts engaged in discussions about the technical hurdles relating to CCS. Subsequently, participation in the ZEP network enabled them to take the step from technology R&D activities to development of policy proposals. Given their technological background, it was only logical that these R&D actors should propose technology-development measures, rather than market-flavoured measures that emphasized only the most economically robust CCS projects. After CCS had become a policy issue within the EU, the European oil corporations decided to support this more actively. This fitted well with their tradition of seeking increased environmental legitimacy and their historical ties to environmental foundations such as Bellona. However, it was not in line with the market logic then dominant within the field – the CCS experts were engineers, and thus not as marked by the market logic as the corporate headquarters of the petroleum industry.

The utilities were not proactive policy promoters to the same extent as was the oil industry, which is understandable in light of the different institutional character and traditions of the two industries. First, the R&D tradition of the electricity industry did not promote CCS: there was no prior engagement in enhanced oil recovery, or R&D units involved in CCS-relevant work. Second, the utilities lacked a tradition of seeking increased environmental legitimacy. Instead, they aimed to secure their coal interests in other ways than engagement in developing climate mitigation efforts. Interestingly, it seems that it was the EU CCS policy process that eventually led the electricity industry to embrace CCS as a CO_2 abatement option. The societal processes relating to policy development made this industry change its interest perception. In essence, the fact that the CCS idea stemmed from the R&D departments of the oil industry created what can be seen as institutional feedback that gave the EU CCS policy a technology-development character.

The CCS financial mechanisms, which centralized the competence distribution of this policy issue, are not as deeply rooted in the European organizational field of petroleum production. It was individuals from Shell, BP, Alstom, Vattenfall and environmental foundations who promoted this outcome. In order to understand why they succeeded, we need to take the entrepreneurial character of their efforts into account.

We should also note that NI lenses can sensitize us to the lack of issue-specific institutional traditions at the national level; no countries had longstanding national CCS policy traditions. Because the member states did not involve themselves in defending existing regulatory traditions, it was easier for the oil industry and other CCS entrepreneurs to gain acceptance for their policy proposals. Moreover, this lack of prior national policies made the member-state governments more independent of their national industries than the case in policy areas where governments and industries are embedded in common regulatory traditions. Lack of EU policy traditions in the policy area contributed in much the same way. True, DG Environment officials feared that a too-strong CCS policy reliant on technology-development criteria might threaten the dynamics and functioning of the ETS. But these officials did not have the time or capacity to involve themselves heavily in CCS policy development after 2005.

The only institutional feedback at work here is found within the European oil industry. As to the character of the main EU CCS policy outcome, the NI perspective helps to shed light on the steering method chosen. The technology-development character of the policy clearly reflects the institutional logic of the experts involved in making this policy – mainly people from the research units of the oil industry. However, the brief history of European CCS policy makes institutional feedback mechanisms less prominent here than in other cases examined in this book. On the other hand, it does seem as if the absence of institutional traditions in the policy area created more space for entrepreneurship. To understand the significant centralized control that came to characterize the policy outcome, we need to look into the entrepreneurial activity in this policy area.

In order to highlight the network entrepreneurship mechanism developed on the basis of the MLG approach we will track the actors who exhibited high intensity or creativity in the policy-development process. We find little during the first period: prior to 2005, no actors seem to have invested much time or energy in the development of an EU CCS policy – but this soon changed. In 2005 and 2006, Commission officials, particularly within DG Environment and DG Energy, brought the CCS issue to the fore, placing it on the EU climate policy agenda.

As we have seen, this was not a result of demands from powerful member-state governments or industry; instead, Commission officials used their formal powers in a creative manner. The Commission initiated a whole range of CCS-devoted forums, including a specific working group under the second Climate Change Programme and the ZEP platform. These forums enabled the Commission to bring many kinds of actors aboard the 'CCS train'; ZEP, for instance, was instrumental in encouraging technology experts from the oil industry to come forward with policy proposals.

The CCS agenda-setting performed by the Commission can be seen as an entrepreneurial act in itself, while the creation of the various CCS forums provided platforms for further entrepreneurial activities. However, the massive workload relating to climate policy in general from 2006 to 2009 made it hard for the Commission itself to exploit these opportunities. Moreover, important

Commission officials remained sceptical to the emission performance standard as well as the financial mechanism.

Instead, other actors exploited the foundation for multi-level networking created by the Commission – especially as regards the financial mechanism and emission performance standard issue, and less so with the development of the 2009 Storage Directive. The latter was developed chiefly through dialogue between the Commission and the ZEP platform, with the focus on resolving technical issues. Entrepreneurship was not so important here: the process was characterized by expert deliberations and technical clarifications.

Entrepreneurial activity was substantially higher with the proposed emission performance standard and the financial mechanism. The Parliament rapporteurs for CCS and ETS were particularly important as entrepreneurs, together with some environmental foundations and a few industry organizations. All these actors engaged far more than strictly 'required', deliberately seeking to make the most out of the situation. Our interviews indicate that these actors felt that the considerable attention given to other issues in the climate package provided more room for entrepreneurship as regards CCS.

Furthermore, they had a tactical 'stick and carrot' plan behind the combination of the two proposals: the emission performance standard was a negative 'stick' towards the polluting industries which could make it easier to muster support from Green MEPs for the financial mechanism. The latter was then more of a 'carrot', rewarding large source polluters that invested in CCS. The entrepreneurs included renewables in their proposal in order to gain support from MEPs as well as member states (and we note that Spain followed up on this enthusiastically in the very final phase of climate package negotiations). Moreover, the entrepreneurs tactically shifted their line of argument according to whom they talked to, in order to get more people on board.

As further elaborated in the subsequent interaction analysis, the concurrent deliberation of the ETS revision and the CCS process gave entrepreneurs a unique opportunity to develop a financial mechanism and strengthen the EU CCS policy, its centralized character in particular. However, this linkage between the two issue-areas was not the result of a deliberate plan from the Commission or anyone else, but should be seen as a fortuitous situation that the entrepreneurs skilfully exploited. We should also note that time-pressures related to the need to finalize the package provided Parliament rapporteurs with greater leeway than in more regular processes, forcing the Council to develop compromises. Several of our interviewees indicate that the entrepreneurial success hinged on the special procedure for finalization and adoption of the climate package.

This entrepreneurship had a truly multi-level character. Its actors were primarily Brussels-based, and reached out to actors at other societal levels, most importantly by touring the member states. In addition to creating networks and exploiting the window of opportunity, these entrepreneurs also contributed to shaping how powerful industries perceived their self-interests – a prime example being how the electricity industry eventually came to see an EU CCS policy as important for its own future survival.

The European electricity industry is heavily reliant on coal, and since this will be a challenge under a tougher international climate regime, this industry may be inclined to embrace CCS as a carbon abatement option. Eventually, these arguments did come to influence how the industry perceived its self- interests, but this appears to have been more the result of the EU policy development than a reason for policy adoption. The entrepreneurs contributed to change how the industry perceived their interests in a three-stage process: first, the argument was crafted by non-industry actors in Brussels (MEPs and non-state actors); second, the entrepreneurs persuaded member states that this was the best way of understanding industry interests; and third, the industry came to adopt this understanding of its own interests.

On the whole, we find the network entrepreneurship mechanism helpful in understanding the CCS policy development process, and how industry acted to shape the outcome. To a significant extent, industry actors acted as highlighted by this mechanism. But the efforts of MEPs were equally important, with the Commission playing more of an enabling role. We also find a different kind of entrepreneurship than networking and exploitation of windows of opportunity. The entrepreneurs were able to change how core actors – the electricity industry in particular – understood their own best interests. This entrepreneurship took more the form of influencing the normative views of industry actors.

It is clear that industry did not provide the hub in the CCS network. The Commission was important for creating the central window of opportunity, but it did not coordinate its efforts closely with the other actors. It was the MEPs, the CCS-promoting industry and environmental foundation actors who coordinated their efforts more closely. Here we note that with industry, there were more actors jumping onto the bandwagon than setting it in motion. The entrepreneurial success in this case was partly a result of the skilled and professional conduct of the entrepreneurs. Let us now turn more specifically to the question of internal policy interaction, further clarifying the linking activities of entrepreneurs.

Policy Interaction: Important Bargained Interaction with the ETS

Since the EU CCS drive came about as a part of the broader EU climate policy drive from 2005 on, we may expect it to have been more profoundly shaped by interaction than the other cases. CCS is not a policy with a long institutional history in the EU (such as renewables), nor was it established as a cornerstone policy meant from the start to influence other policy areas (as in the case of emissions trading). These characteristics alone indicate that we could be talking about a policy destined to be more of a receiver than a sender.

Which policy interaction mechanisms were in operation in the CCS case? Using functional interaction lenses, we might argue that the volatile and generally low carbon price created a need for an additional and more specific mechanism for funding CCS. However, it was not obvious that this need should be met; it could also be argued that the lack of a stable and high carbon price made CCS less attractive as a climate measure and that functional interaction hence undermined this mitigation technology. Still, the prospects of ETS auctioning revenues, and the various measures negotiated at the same time within the climate package, formed a crucial and 'handy' basis for the interaction that was to unfold in the course of 2008.

As regards bargained interaction, we see that some skilled entrepreneurs were able to exploit this situation. Civil society actors came up with the idea of a more elaborate and specific economic incentive link between the revised ETS and the construction of CCS demonstration projects. The two Parliament rapporteurs for CCS and the ETS, respectively, carried out the actual linkage. With their skilled tactics, the two rapporteurs succeeded in mustering support for this issue linkage. The collaboration between Chris Davies and Avril Doyle is certainly remarkable, particularly when we recall that they represented different party groups. The issue linkage was also facilitated by the fact that the two pieces of legislation were parts of the same package, although neither the Commission nor member states had originally planned for such a connection. This linkage should be seen as a lucky situation that the entrepreneurs managed to exploit skilfully.

Liberal Intergovernmentalism argues that member-state representatives can use issue linkages to promote their cause. However, in the case of CCS it was not member-state representatives but MEPs that created the issue linkage. Member-state officials assumed a more responsive role, accepting the linkage created by the Parliament and negotiating the precise ETS part of CCS funding. The UK was heavily involved in garnering support from other actors, and ensuring that the ultimate CCS deal could be reached in the very last minutes of the meeting. However, the British seem to have based their tough bargaining more on the general weight and importance of the UK in EU policymaking than on an issue-linkage tactic of being flexible in other issue-areas to ensure the preferred CCS outcome.

Was there then some important institutional interaction at play? Not very specifically or directly, it would seem. The NER 300 fund is a special construction that gives the EU competence to steer national industry development in a way not evident in other policy areas. True, the policy outcome as regards renewable energy also had an EU engineering quality, but in that case the EU did its steering more through binding targets and technical criteria, not direct support to technology-development projects (see Chapter 5).

Further, we note that the market design of the ETS did not really affect the steering method of the CCS policy, despite the tight linkages created between the two policies. Thus we conclude that the notable degree of centralized competence in the CCS policy outcome was shaped to a significant degree by bargained interaction with the ETS, to a certain extent underpinned by functional interaction. As we will see, the EU-external environment contributed further to strengthen policy centralization here.

External Environment: Enhancing the Effect of Entrepreneurship

The EU began to develop a CCS policy only after CCS had emerged as an issue in various international forums, and the international energy security situation had become more challenging. This chronology would indicate that international developments led to the EU-internal shift – but is it really tenable to argue that the EU-internal developments occurred as a direct response to changes in the external environment?

Let us examine our perspectives and mechanisms in turn. We begin by asking whether international developments brought coercive pressure to bear. As regards the energy security challenge, with developments like soaring energy prices and unsecure Russian gas deliveries, this may have led EU officials and member states to search for technologies to boost the use of domestic sources of energy. But we have not found empirical evidence to indicate that this had any direct effect on the emergence of a CCS policy in the EU. It was possible to respond to changes in conditions on the international energy market by a whole range of measures, of which CCS was only one.

Nor was there anything in the energy security challenge as such to indicate that technology development and strong centralization would be appropriate characteristics of EU policy responses. As for international CCS developments more specifically, international CCS decisions and developments (like the US FutureGen programme and the G8 decision in 2008) did not require the EU to develop a CCS policy with certain characteristics.

Could it be argued that the EU engineering policy development in this issue-area was based on international policies that promoted and diffused the technological development of CCS? Before 2005, CCS was promoted primarily by the USA, but this proved more of a problem than an enabling factor for the Commission officials who wished to include CCS as an important EU climate policy issue. After 2005, the increasing involvement of more broadly credible, international research-based organizations like the IPCC, IEA and OSPAR helped CCS to appear as a legitimate and promising technology for climate-change mitigation. Moreover, international initiatives had a technology-development focus: the view that it was a goal in itself to develop CCS, not to focus solely on low-cost abatement options. Hence, it seems reasonable to argue that international developments fostered the development of an EU CCS policy.

However, none of the developments noted above affected the EU policy processes *directly,* although they made it easier for EU-internal entrepreneurs to argue for EU-specific CCS policy development. True, the OSPAR decisions relating to CCS were important in facilitating the Commission's drafting of the Storage Directive, but the Commission might well have opted for a technology-development focus in any case. Hence, we conclude that the institutional diffusion mechanism had only a limited direct impact on the development of EU policy on CCS.

Turning to international entrepreneurship, we focus on actors that actively coupled the EU policymaking to other international forums in order to strengthen their clout in the EU's international policymaking. The UK and to some extent Germany were active in international processes that discussed and promoted CCS, but we have no evidence indicating that Germany and the UK actually pushed for international progress in order to subsequently step up the pressure for internal EU policy development. However, as discussed in earlier chapters, the Commission and core member states (particularly Germany and the UK) argued that the EU needed credible policies to underpin its target of 20 (30) per cent reductions of EU greenhouse gas emissions. The rationale was to strengthen the EU's position as the global climate leader. These actors used the approaching Copenhagen summit to nurture a shared feeling of urgency, and build agreement on replacing the regular formal procedure with an exceptionally swift process.

As discussed in Chapter 3, developments within the climate regime created a window of opportunity for new EU climate policy initiatives. This gave the Commission room to initiate an EU CCS policy development in 2007, as well as providing CCS entrepreneurs with important backing when they initiated the CCS financing mechanism linked to the ETS in 2008. The window of opportunity created by the Commission and core member states strengthened the credibility of arguments that CCS was necessary for tackling future climate obligations and challenges. This kind of entrepreneurship, involving the active use of international developments, has arguably an institutional character that is not quite captured by the MLG way of describing entrepreneurship.

On this backdrop, we conclude that international entrepreneurship, as highlighted by MLG, seems to be the essential thing to note here. True, there was institutional influence from the global climate regime after 2005, with the increasing involvement of organizations like the IPCC and IEA helping CCS to appear as a more legitimate and promising technology for climate-change mitigation. However, the international institutional developments paved the way for entrepreneurship, rather than affecting the EU policy development processes directly.

Conclusion: The Success of *Carpe Diem* Entrepreneurship

In this chapter we have explored why EU CCS policy came into being and why it ended with an 'EU engineering' outcome. Industry had a minor impact on the emergence of this policy, but more influence on its character. Policy interaction played a major role: the link to ETS explains why the policy ended up as a highly centralized policy outcome. The external environment was important in that it helped to create a window of opportunity, and helped to legitimize CCS as a policy option.

Specifying the EU-internal dynamics, and contrary to Liberal Intergovernmentalist claims, it seems that factors and forces other than the influence and pressure of domestic industry were most influential in shaping member-state positions. Arguments relating to electricity industry interests did induce member states to change their positions during the policy development processes, but these arguments were not rooted within industry itself.

European CCS policy has only a short history, and that has made institutional feedback mechanisms less prominent. The only long-term institutional developments that underpin policy development here are found within the European oil industry: the steering method was flavoured by the technology-development logic of the oil industry engineers who came up with the initial policy proposals.

The network entrepreneurship mechanism, extracted from the MLG approach, has proven helpful in understanding the process of CCS policy development, and the role of industry. To some extent, industry acted as highlighted by this mechanism: they established networks, they sought out and utilized windows of opportunity. In addition, the efforts of MEPs were particularly important. The CCS entrepreneurs, both industry and MEPs, exploited the feeling of urgency surrounding the development of the climate package and the special political procedure of the package to the maximum. Thus we call this *'carpe diem* entrepreneurship' – as opposed to the persistent and lengthy 'tortoise entrepreneurship' witnessed in the case of the ETS.

By highlighting policy interaction we have further clarified the linking activities of entrepreneurs. To a significant degree, the policy outcome was shaped by bargained interaction, with the European Parliament as the focal point. In particular, the link between the ETS and CCS was facilitated and developed by the two Parliament Rapporteurs for CCS and the ETS, respectively. In addition, the EU-external environment had important indirect effects on the internal policy development: first, the situation in the climate negotiations enabled the Commission to create a window of opportunity for climate policy development; and, second, the increased CCS focus in the international climate negotiations and G8 made it possible for the Commission to launch a CCS directive. It was the international institutional entrepreneurship of the Commission that paved the way for the subsequent *'carpe diem* entrepreneurship' of industry, MEPs and environmental organizations.

Chapter 7

EU Energy Policy for Buildings: A Result of Failed Entrepreneurship?

Introduction

For decades, the energy performance of buildings has been central in EU energy and climate discussions. The European building stock represents a major obstacle to the creation of a low-carbon energy system: it consumes 40 per cent of European energy and is responsible for about half of the CO_2 emissions not covered by the EU Emissions Trading System, the ETS (European Commission 2008i: 3). Hence, enhanced energy quality of buildings will lead to mitigation of carbon emissions as well as increasing the share of renewable energy.

Despite all the talk, the actual policy result has been meagre: the EU energy policy for buildings is not as centralized and binding as we see in other climate policy areas. In 2010, the EU adopted the Energy Performance of Buildings Directive (or EPBD, European Parliament and Council 2010), which has what we would call a 'local loading' character. It requires member states to develop various kinds of measures to promote technology development, but the national governments have the competence to shape the actual character and ambitions of these measures.

The Commission has tried to act as an entrepreneur for improved energy performance of buildings, but without the same success as in other areas of climate policy. Repeatedly, the Commission has pointed out that transformation of the building stock is central to EU climate strategies, but the member states have remained sceptical to centralization here. Furthermore, it is not clear what is meant by 'the energy performance of buildings', as different actors give varying weight to three main components of a building.

Some aim to enhance the thermal quality of the building envelope (improved insulation, preventing energy leakage); others focus on the efficiency of the internal distribution of energy; and finally, one can argue that installation of low-scale renewable energy-producing onsite is the key point (European Commission 2009c, Jensen, Wittchen and Thomsen 2009). Moreover, definitions of 'high energy-performance buildings' flourish, with this new concept going under many names – like 'passive houses', 'low-energy buildings', 'fossil-fuel independent houses' or 'zero-energy buildings' (European Commission 2009c, Schild, Klinski and Grini 2010: 22).

This ambiguity has made it hard to develop a substantial, coherent EU policy for buildings, although a tangible local loading policy has eventually emerged.

The main questions addressed in this chapter are: Why has the EU developed an energy policy for buildings? Why does this policy have a 'local loading' character? Here we will explore how industry, interaction and EU-external mechanisms have affected the development of policy.

This case contains elements relevant to all the three themes in focus in this book. First, there seems to have been relatively little involvement on the part of industry. Achieving better energy performance of buildings requires the construction of new buildings with 'more' when it comes to building components – more insulation, more technical equipment, windows with more glass, etc. Despite this apparently 'lucrative' backdrop, few industry actors showed much interest or engagement in the EU policy processes. Why this relatively modest involvement on the part of industry? And did this lack of industry engagement create a power vacuum that could be exploited by other actors?

Second, the EU energy policy for buildings started out as an energy-efficiency policy but ended up promoting renewable energy as well. Moreover, both EU policy areas rely on technology-development steering and have often been discussed in parallel political decision-making situations. Hence, we have reason to expect to find interaction. To what extent and how was the 2010 Directive shaped through interaction with renewable energy?

Third, EU policy for buildings has been strengthened three times, all coinciding with critical developments in international climate negotiations (in the early 1990s, at the entrance to the 2000s, and in 2008–2009). Since the climate regime can hardly be said to feature any specific advice with respect to energy efficiency, it is natural to explore how these two spheres of policymaking may have interacted. What is the actual causal linkage between the global regime and the EU policy development? And how important has this been for the actual outcome? As in previous chapters, we draw on three approaches –Liberal Intergovernmentalism (LI), New Institutionalism (NI) and Multi-level Governance (MLG) – in seeking to specify the mechanisms through which industry, interaction and the external environment have affected the development of policy.

The Outcome: Local Loading

The EPBD aims to reduce energy consumption and to increase the use of renewable energy in the building sector (Directive 2010/317/EU). It contains five main elements: a foundation for a holistic method for calculating the energy performance of buildings; requirements directed at national building codes; obligations on the regular control of heating and cooling systems; recommendations for increased use of national financial support measures; and a general outline for national energy certification of buildings.

'Technology-development' steering is dominant. The method for energy calculation provides the basis for all other measures. This is a holistic approach that takes into consideration both the thermal quality of the building envelope and the use of renewable energy. The method is focused on primary energy (Art. 2). This means that the *absolute* energy use of the building is what counts: it is not possible to compensate for building envelope with poor thermal quality by taking into account on-site technical equipment. A key concept here is 'cost optimal', defined as 'the energy performance level which leads to the lowest cost during the estimated economic lifecycle' (Art. 2.14). In contrast to cost-efficiency considerations, cost-optimal calculations are to take the whole lifetime of the building into account. Such longer-term calculations will make costly measures appear more economically viable.

The EPBD requires member states to apply minimum requirements to new buildings and building units, existing buildings and building elements that are subject to major renovation, and to technical building systems whenever they are installed, replaced or upgraded (Art. 1). The member states cannot choose to regulate only some parts of the building (like the thermal quality of floors and walls): the holistic energy calculation method must be applied. Minimum requirements are to be reviewed at least every five years, providing an incentive to regular strengthening of the requirements (Art. 4). Further, member states are to choose cost-optimal minimum levels. In addition, all building elements with significant energy impact that are retrofitted or replaced shall be replaced with elements that meet minimum energy requirements (Art. 7). Member states must also inspect heating and air-conditioning systems and buildings, and independent control systems are required (Articles 14, 15, 17).

By 2020, all new buildings are to be nearly zero-energy; public buildings must fulfil this requirement by 2018 (Art. 9). Member states must develop national plans for 'nearly zero-energy buildings', and 'take measures such as the setting of targets in order to stimulate the transformation of buildings that are refurbished into nearly zero-energy buildings'. Interim targets are to be in place by 2015. Member states are encouraged to introduce financial incentives in order to catalyse the energy performance (Art. 10). Every three years, member states shall draw up lists of existing and planned financial instruments. Based on the input from member states, the Commission is then to prepare report on the use of EU funds and, if it finds this appropriate, submit proposals to the European Parliament and the Council.

In contrast to the other elements in the EPBD, the energy performance certificates have a market flair. These certificates shall indicate the energy performance of a building or a building unit, and be calculated in accordance with the methodological requirements of the Directive (Art. 11). Member states are to issue certificates for all buildings or building units that are constructed, sold or rented out, or have useful floor area of over 500 m² (to be lowered to 250 m² in 2015) (Art. 12). The certificate shall figure in building advertisements in commercial media, and all buildings that are frequently visited by the public must display the certificate in a prominent place.

Independent control systems are mandatory, in order to ensure the credibility of these certificates (Art. 18). In addition, the Commission shall by 2011 adopt a voluntary, common EU certification scheme for non-residential buildings (Art. 11). Similarly, the certificates may affect the market value of buildings. The certificates are also to include recommendations for 'cost-optimal or cost-effective' improvement. Hence, energy certification is not only a market measure: it also provides people with technical guidance on how to enhance the energy performance of their buildings.

As the EPBD lacks clear, binding commitments, it does not transfer much authority to the EU. It merely gives the Commission competence to set a common comparative methodology framework for calculating the integrated energy performance of buildings and building units, and to monitor and become involved in various national processes, through monitoring their follow-up measures (like their zero-energy planning), and through regular inspection of heating and air-conditioning systems. It is specifically stated that it 'is the sole responsibility of Member States to set minimum requirements for the energy performance of buildings and building units' (Preamble 10).

Core concepts like 'nearly zero energy' and 'cost optimal' are unclear. A draft issued in May 2011 indicates that member states will have considerable leeway to develop 'cost-optimal energy calculation methods' in accordance with their own preferences (European Commission 2011b). The Directive's definition of a 'nearly zero energy building' is highly ambiguous: a 'building that has a very high energy performance' and its 'nearly zero or very low amount of energy required should be covered to a very significant extent by renewable energy sources' (Art. 2). Member states are free to interpret what this is to mean in practice.

Thus, we may conclude that the EU energy policy for buildings is a fairly clear example of what we have termed a 'local loading' policy. Although the EPBD does contain one market-oriented measure – the certificates – it is otherwise based on a technology-development steering method. Moreover, member states are given leeway to determine whether the energy certificate schemes are to be designed so as to affect the market price, or whether to use this instrument in order to give guidance to the public on how to improve the technical quality of their buildings. In addition, member states also have considerable leeway in determining how to effectuate the other parts of the Directive. The EPBD contains no targets or timetables: actual ambitions are to be determined by the national governments.

The roots of this policy development go back to the early 1970s. Let us now present the story.

The Story: From Mere Symbolism to Local Loading Policy

The 1970s to 1997: The Energy Efficiency of Buildings becomes a Political Issue

In the immediate postwar years, building construction gained political attention as part of national welfare policies; but as the quality of the building stock improved, this issue left the political limelight (Kemeny 2001). It was only later that buildings come to attract new political attention by virtue of their 'energy performance'. This re-conceptualization of buildings can be traced back to the oil crisis in the 1970s, when most West European countries adopted energy efficiency standards for new buildings, as well as RD&D programmes, information and educational efforts and financial support for energy-efficiency upgrading of homes (European Commission 1979, Geller et al. 2005). Denmark was at the forefront, developing strict building codes and offering lucrative energy-efficiency financing (European Commission 1984a: 5, 20, European Commission 1998a). Because the East European countries relied on imports from Russia, they were not affected by the Arab oil shock, so we do not see a parallel early peak in energy efficiency attention there (Urge-Vorsatz, Miladinova and Paizs 2006).

From the mid-1970s, the EU prudently started to engage in energy efficiency (European Commission 1975, 1979). This was stepped up in 1984, when the Commission launched a holistic energy-efficiency strategy for the building sector, arguing that this sector was responsible for some 40 per cent of the Community's energy demand (European Commission 1984b). However, the member states rejected the Commissions' proposals for a 'thermal auditing' system modelled on the Danish practice, and a common reference building code for the whole Community. Nonetheless, the Council adopted a new overarching objective: 20 per cent improvement in energy efficiency by 1995 (European Commission 1987a). This figure was similar to the energy-efficiency improvement achieved in the preceding decade (European Commission 1984a: 3).

However, fears of energy shortage had long since disappeared by the mid-1980s. Denmark continued to launch ambitious new policies, but most member states reduced or abolished their energy-efficiency programmes (European Commission 1987a, 1990, 1998a). In addition, many de-regulated their national housing markets, thereby removing various kinds of detailed technical energy regulations.

In the run-up to the Rio Summit in 1992, the Commission launched a new energy efficiency offensive, with explicit reference to the EU's increasing energy dependence, the global climate ambitions of the EU, and the EU target of stabilizing emissions at 1990 levels (European Commission 1990; 1992b). As mentioned in Chapter 3, the SAVE programme was launched, offering financial project support and networks for coordination.

Moreover, the Commission proposed a new draft directive, encouraging member states to enhance thermal insulation and to introduce regular inspection of boilers (European Commission 1992c). The draft also re-named 'thermal auditing' as 'energy certification of buildings'. This instrument aimed at providing buyers and lenders with new information, thus affecting market-price mechanisms. According to an interview with a former EU official, the introduction of market thinking reflected the current political climate in Brussels. At the same time, a voluntary labelling scheme for buildings began in the USA, but the full-fledged 'energy star' label did not come until a few years later (Energy Star 2011). None of our interviewees saw these developments in the USA as important to Commission thinking at the time.

The Parliament endorsed the Commission proposals and called for higher ambitions (European Commission 1993). As a result, the Council adopted a 'directive to limit growth in carbon dioxide emissions by improving energy efficiency', the SAVE Directive (European Council 1993). This Directive included all the elements in the Commission's proposal, but the wording was ambiguous; moreover, aside from an annual reporting commitment, it did not place any specific obligations on the member states.

By the mid-1990s, most European countries had included energy requirements in their building codes. Piecemeal and modest approaches to energy-efficient buildings reigned: only the thermal qualities of certain parts of a building – like roof, walls and windows – were regulated (European Commission 1984a: 8, Visier, Thomsen and Johanssen 2003: 7). Despite similarities in the central position of building codes, we can distinguish three groups of countries in terms of their energy policy for buildings. First, in addition to Denmark, both Sweden and Germany had developed ambitious building codes and extensive public funding (European Commission 1998a, EuroAce 2010). Second, many other member states cut back on their spending, whereas others had never really had ambitious policies. Third, the East European countries hardly focused on energy efficiency at all, but since they were non-members at the time, this did not affect development of EU policy.

We may conclude that the energy efficiency of buildings became a political issue in this period. While the 'leader countries' developed technology-development steering, the Commission tried to induce a new market instrument: energy certificates. However, EU policy was very weak, primarily of symbolic importance, and in the early 1990s most member states scaled back their energy-efficiency policies.

1997–2004: From Piecemeal to Holistic Approach

By the mid-1990s, technological research communities had become involved in developing various methods for improving the energy performance of buildings (Keulenaer and Gerwen 2006, Jensen, Wittchen and Thomsen 2009). These experts did not focus solely on the energy character of individual building components, but aimed at improving the total energy performance (Jensen, Wittchen and Thomsen 2009: 12–13). Reduction of energy consumption was highlighted as the main overarching design criterion, with shifts to renewable energy and improved efficiency of fossil fuels as secondary concerns (Keulenaer and Gerwen 2006). The German 'passive house' became a particularly well-known version of this new approach (European Commission 2009c). Thanks to the use of solid insulation, passive solar heating and coherent energy planning, such buildings need hardly any heating or cooling.

By the late 1990s, the European Commission was repeatedly arguing that the EU's Kyoto commitment taken on in 1997 created a need for a more stringent energy efficiency policy (see European Commission 1998a, European Commission

2000c) – particularly since the 1995 target was beyond reach, with only 12 per cent and not 20 per cent energy efficiency improvement. The Commission presented the building stock as Europe's largest energy demander and CO_2 emitter and it linked energy efficiency with the enhanced use of renewable energy in buildings (European Commission 1998a). The Council agreed that a more ambitious EU policy was needed – and so DG Energy could start work on drafting a new directive.

In parallel, DG Enterprise created a working group for sustainable construction in 1999, with representatives from industry and member states (European Commission 2001b). This group recommended stronger EU pressure for the creation of national certificate schemes and the introduction of energy improvements in rehabilitation projects. Our interviews with EU officials as well as civil servants indicate that this gave backing to the DG Energy officials who were seeking to strengthen the requirements of the SAVE Directive. Although industry actors played a certain role in the DG Enterprise working group, they were rather passive otherwise, according to our interviewee.

The building construction industry can be divided into three groups: constructors, building product producers, and building managers. As to the first group, the constructors: here we find architects, construction contractors, plumbers, carpenter, roofers, etc. It is a very heterogeneous group, as each of the different professions tends to specialize, with most employed in specialized firms that represent one profession only. The construction industry is the largest industry employer in Europe, but most firms are small and local – the European Construction Industry Federation (FIEC) estimates that 95 per cent of the construction firms in the EU area have fewer than 20 employees (FIEC 2010). Not many companies are publicly traded. Our interviews indicate that few of these actors follow strict market logic: firms generally focus on surviving in the business and operate with fairly short planning horizons (see also Boasson 2011).

Constructors are also characterized by deep national differences (Schild, Klinski and Grini 2010). One interviewee put it like this: 'As all the European countries have their own kitchen, like the French and the Italian cuisine, they also have their own characteristic construction businesses.' These actors are typically represented in Brussels by small business associations that again consist of a range of national associations. Our interviews indicate that few corporations from within this group lobbied the EPBD process actively, and that FIEC assumed a reactive and not proactive role (see also *ENDS* 2000d). It should also be noted several large transnational construction contractors (like German Hochtief, French Vinci Construction and Swedish Skanska) had no direct representation in Brussels.

In the second group, the building product producers, we find industrial actors like cement, steel, flat glass and insulation. Each sub-industry here is dominated by fairly concentrated industries, comprising five to fifteen large enterprises with a European or global outreach (see e.g. Nordqvist, Boyd and Klee 2002, Eurima 2011, Glass for Europe 2011). Most firms are highly professional and are publicly traded. Our civil-society interviewees indicated that, because of the relatively low number of firms, Eurofederations could develop clear preferences.

However, whereas the insulation and glass industries create products that are applied only or primarily in buildings, many of the others make products used for many purposes. That means that few product producers were involved in the EPBD processes to any extent, with the exception of Eurima, the insulation trade association. It supported energy certificates and called for stricter insulation regulations for new buildings as well as existing building stock (*ENDS* 2000e). In addition, the insulation industry initiated the creation of EuroACE, an organization that promotes a more ambitious EU energy policy for buildings (EuroACE 2011). Our interviewees among EU officials and civil-society actors confirmed that the insulation industry acted as eager lobbyists.

Third, the building managers. They consist of a wide range of actors – commercial landlords of various sizes, firms and organizations that ensure public, cooperative and social housing, and of course all the individuals and small/medium firms that own their own house or buildings. Real estate became a financial object in the 1980s, but professional owners did not get Brussels presence until the European Property Federation was established in 1997 (EPF 2011). According to our interviewees, building managers did not engage much in the EPBD processes, except from expressing doubts as to energy certification. Public, cooperative and social housing organizations are represented through a common organization, CECODHAS (now Housing Europe). Our interviews showed that this organization was chiefly interested in getting tenant rents reduced, and was hardly involved in the first EPBD process. Not surprisingly, individual owners are not represented in Brussels.

We may conclude that the building construction industry is fragmented. As one interviewee put it:

> What is very fascinating in this industry is that all of the actors involved are so focused on their task. They do not grasp the totality, that they are actually building a coherent building. If they work with insulation, well then that is where their focus will be (...) You must recall that this is new territory, the building has never been regarded as a coherent entity before. Now the actors have the change their perceptions!

Due to the special characteristics of this industry, the Commission was not able to have an extensive dialogue, with the exception of representatives of the insulation industry. Hence the draft directive was primarily shaped by the Commission, taking further and specifying core features of the SAVE Directive (*ENDS* 2001c). In addition, the Commission was inspired by the recent development among researchers, and proposed an outline of a more holistic calculation of energy use in buildings – and here the Commission proposed that the use of renewable energy be taken into account. Most notably, the key term in the draft, 'the energy performance of buildings' was a new invention that reflected the cognitive shift among experts. The Commission's EPBD draft identified three components as crucial: energy, minimum national energy requirements for buildings of more than 1500 m^2, and regular control of boilers.

It was by no means full agreement on the draft in the Commission. On the very day it was issued, the college of commissioners requested a further redrafting, arguing that the matter belonged under national sovereignty and should fall under the subsidiary principle (*ENDS* 2001c). This was an unusual move, not least because the Commission draft already offered member states substantial leeway (*ENDS* 2001c). As noted by one interviewee: 'Subsidiarity was a huge battle. This was the prerogative of the member states'. Most member states were reluctant to delegate authority to the EU level. The three countries that had already developed the most specific and comprehensive national policies (Denmark, Germany and Sweden) were positive to the Directive, but they did not campaign for greater centralization of control.

Denmark was the most active; the country also provided inspiration for core elements in the EPBD. Here we may note that the insulation industry has a particularly strong position in Denmark; two of the leading global insulation corporations have their base there (Isover 2011, Rockwool 2011). Germany and Sweden tightened their national regulations at the time the EPBD was developed (*ENDS* 2001a, Fuglseth 2009). Germany introduced changes in its regulations with regard to the application of primary energy calculations, a move that promoted the use of renewable energy and not only energy efficiency (*ENDS* 2001a, 2002a).

Despite this seemingly substantial member-state reluctance, the EU energy ministers reached informal agreement on the directive after half a year, proposing only a few changes. However, the ministers did agree to postpone the implementation deadline until 2009 (five years after that proposed by the Commission), limiting energy requirements for building renovations costing more than 25 per cent of the value of the building, and requiring certificate updating only every ten (not five) years (*ENDS* 2001d). EuroACE and environmental organizations argued that the later entry into force 'knocked out' the potential of the Directive to contribute to the EU effort to meet its Kyoto targets (*ENDS* 2001d, 2001e). This was supported by the Parliament Rapporteur, who stated that the Council position was 'illogical' since it would allow countries to delay implementing the Directive until after the start of the Kyoto commitment period (*ENDS* 2002b).

After a few months of deliberations, the Council and the Parliament agreed on a compromise EPBD, requiring member states to implement the directive by 2006, but this could be postponed to 2009 if they lacked qualified/and or accredited experts (*ENDS* 2002c, European Parliament and Council 2002: Art 15). This swift process was possible, according to our interviewees, because few member states paid much attention to the political deliberations on the new directive. While much of the policy discussion related to the implementation deadline, the Council and the Parliament also contributed to change other details of the Commission draft.

The obligation to include energy requirements in national building codes was strengthened slightly: for instance it was to apply for all buildings larger than 1000 m², not 1500 m² as originally proposed by the Commission (European Parliament and Council: Art 3, 5). The regulations were to be reviewed (implicitly implying strengthening) at regular intervals of maximum five years (Art.4). Further, energy certification, should apply to new as well as existing buildings (Art. 7, 10). It was indicated that energy certificates should be made available when buildings were constructed, sold or rented out, and that buildings larger than 1000 m² should display the energy certificate clearly visible to the public. The intention was that information on the energy performance of buildings should affect the price mechanisms in the building market, but the wording of certification was somewhat ambiguous and open to interpretation.

Hence, the EPBD emerged with what we call a 'local loading' character: little power was delegated to the EU organizations, and there was scant pressure towards the development of national market instruments. Most other parts of the Directive left it to the member states abilities to require technical improvements of the building stock. Member states could choose to develop certificate schemes that affected the market pricing of buildings, but they were also free to design certificate systems that promoted specific technical improvements. As a further step towards developing common templates for building regulations, the Commission later engaged the European Committee for Standardization (CEN) to develop standards to facilitate national implementation. By around 2004, CEN had issued 31 standards relating to EPBD (CEN 2009). This move strengthened the technological flavour of this piece of regulation further, but the member states still retained substantial leeway in interpreting and implementing the templates.

In those years, the holistic approach to the energy performance of buildings spread within the European expert community. Many definitions of 'high energy-performance' buildings emerged, differing in the weight given to on-site renewable energy generation, carbon emissions and thermal qualities (European Commission 2009c, Jensen, Wittchen and Thomsen 2009: 12–13). However, actual construction of such buildings was still a marginal phenomenon: for instance, only a few thousand 'passive houses' were constructed, primarily in Austria and Germany (European Commission 2009c).

The upshot of this is that by 2004, the concept of 'high energy-performance buildings' was about to replace the energy efficiency focus on buildings from the 1980s and 1990s. The EU energy policy for buildings was no longer solely about reducing energy consumption, but also about promoting renewable energy. Moreover, this was an elite project: it was driven mainly by construction experts and the Commission, with little engagement from the huge and diverse building construction industry.

2005–2010: Increasing Ambitions, but Holding onto Local Loading Governance

Energy efficiency gained renewed attention in the EU from 2005 on (European Commission 2005d). The Commission made repeated reference to the preparations for the Copenhagen climate summit, and the soaring energy prices, and concluded that 'energy efficiency is a key part of Europe's response to climate change and security of supply issues' (*ENDS* 2008b). In 2006, the Council embraced a 20 per cent improvement of energy performance target, applicable for the period 2005 to

2020 (European Council 2006: 15). This objective was indicative, not binding. No specific target relating to the energy performance of buildings was developed, and energy efficiency was not included when the EU initiated its ambitious climate and energy package in 2007.

By 2006, only five member states had incorporated the EPBD properly (European Parliament 2008a: 5). The Commission concluded that the 'main contribution of the EPBD, so far, has been in bringing the subject of the energy efficiency of buildings onto political agendas, building codes and to the attention of citizens' (European Commission 2008h: 2). However, nearly all member states had adopted energy-efficiency targets of some kind, although the measurement methods varied so greatly that actual levels of ambitions were hard to compare (European Commission 2009c: 7). In addition, one third of the member states had introduced more holistic energy calculation methodologies in their building codes (EnR 2008: 8). Also here, however, there was great variation in the details of the methodologies and energy requirements (Schild, Klinski and Grini 2010).

A majority of EU member states introduced state aid schemes, most of which followed the technological-development logic (European Commission 2009c: 14–15; Jensen, Wittchen and Thomsen 2009: 47–50). Developments were not quite so impressive with regard to energy certification. This policy recipe did not spread so easily, and by 2009, only a handful of member states had introduced energy certification schemes (EnR 2008: 9, European Commission 2009c: 14). These schemes varied greatly: some worked more as a tool for technical standardization than as a market measure (European Commission 2009c: 13–14, Jensen, Wittchen and Thomsen 2009: 16).

At this stage, the member states were similar in the sense that building codes played a major role, but they were combined with various mixtures of other measures (see e.g. Papadopoulou et al. 2009, EuroACE 2010). Still we can identify three distinct groups of countries, but now the group of frontrunner countries had grown. Interestingly, EPBD implementation led Denmark to strengthen its already strict regulations significantly (Erhorn and Erhorn-Klutting 2009, Thomsen and Aggerholm 2009). However, Danish financial support for energy-saving measures was gradually scaled back due to a shift in the government in 2001. Also Sweden and Germany changed and strengthened their already strong building codes as a result of EPBD implementation (Fuglseth 2009; Schild, Klinski and Grini 2010). Additionally, Germany introduced new requirements to promote on-site renewable energy.

The 'passive house' movement had been gaining a stronger hold in many countries. Austria was particularly successful here, thanks to its many different regional certificate schemes and to its housing subsidy system (Jilek 2010). This in turn meant that Austria now joined the group of 'frontrunners', as did the UK, with its specific focus on CO_2 abatement and on-site renewable energy (Woods 2010). The UK introduced a certificate system that promoted technical change directly, rather than aiming to change the pricing mechanisms of local building markets. For instance, the certificates had a CO_2-based index for describing the energy performance of buildings and detailed proposals for energy-efficiency improvements. Further, the certificates displayed historical energy data and were updated annually. By 2009, all the leading countries had decided that new buildings constructed after 2020 (2030 for Austria) were to be low-energy houses, although these were variously defined and named (Thomsen and Aggerholm 2009, Schild, Klinski and Grini 2010: 22).

The second group of countries included most of the other 'old EU member states' who implemented EPBD more reluctantly (see Papadopoulou et al. 2009). We can take Spain as an example. Implementation of the EPBD was Spain's first step towards strengthening the energy performance of buildings since 1979 (Molina and Álvarez 2009). Certification of new buildings was introduced in 2007 and certification of existing buildings in 2010. Further, Spain's building code promotes renewable energy – for instance, solar energy is mandatory for almost all non-residential buildings and for all domestic hot water. Despite these changes, Spain failed to adopt all parts of the EPBD properly and the Commission issued an infringement procedure against Spanish implementation in 2010 (Build up 2010).

The third group consisted of the East European countries who started to develop energy-efficiency policies after entering the EU. After decades of underinvestment, much of the building stock in these countries was in urgent need of repair and improvement (Peterstorf et al. 2005). Interviewed EU officials pointed out that many of the accession countries liberalized their housing market, but lacked detailed regulation and organization of building ownership. This pulverized responsibility for energy-related investments, and made it hard to develop appropriate regulation.

Poland was no different from the other accession countries; the energy performance of its building stock was poor, and implementation of the EPBD failed to bring much change (Panek and Popiolek 2009). New and more holistic energy requirements were introduced in Poland's building code, but these were open to interpretation and in some respects weaker than earlier regulations. An energy certification scheme was introduced, but with little practical effect, not least because of a special juridical clause stating that a certificate would be issued only if both vendor/landlord and buyer/tenant expressed the desire for this.

Back in 2005, the Commission had issued a Green Paper on energy efficiency that hinted to the possibility of revising and improving the 2002 EPBD (European Commission 2005d: 19). In 2006, the Commission launched an energy-efficiency action plan where it proposed to give the EU some competence in this issue-area, for instance by introducing a common minimum performance standard for new and renovated buildings, and developing a strategy for widespread deployment of very low-energy or passive houses (European Commission 2006d: 12). The European Parliament welcomed the Commission proposals, but called for a greater centralization, for instance suggesting 'a binding requirement that all new buildings needing to be heated and/or cooled be constructed to passive house or equivalent non-residential standards from 2011' (European Parliament 2008c: 9).

The member states were not particularly enthusiastic, emphasizing that the subsidiarity principle should be respected and inviting 'the Commission to take into account the experience from the ongoing implementation of the Buildings Directive before expanding its scope or including minimum performance requirements for new or renovated buildings' (European Council 2006: 4–5). Our interviews with EU officials indicated that lack of member state support and industry pressure made certain Commission officials sceptical to a revision of the EPBD.

Furthermore, market supporters within DG Environment and DG Energy were sceptical to any measures that might interfere with the CO_2 price mechanisms of the ETS. Our interviews showed that the insulation industry pressed for a revision, and this was supported by most DG Energy staff. EU official interviewees confirmed that the recast was launched on the initiative of Energy Commissioner Piebalgs; he was eager to finalize a new directive before his term ended. The media reported that Piebalgs was now supporting energy efficiency 'more enthusiastically than ever' (*ENDS* 2008b).

In the beginning of 2008, the Commission signalled that it was considering the introduction of EU-wide minimum energy requirements (*ENDS* 2008a). The insulation industry, represented by Eurima, enthusiastically supported such centralization of control, but few other industry actors gave any immediate response (*ENDS* 2008d). At this stage, DG Environment shifted its stance and became a promoter of a more thorough revision of the 2002 EPBD (*ENDS* 2008i). However, officials from both DGs soon realized that it this would be a very challenging task (*ENDS* 2008d). Our interviews show that the Commission had dialogues with the insulation industry on developing some kind of procedure that would make the member states improve their performance – without delegating more power to the EU as such. It was eventually agreed that one solution could be to introduce requirements for the application of 'cost-optimal' measures, the rationale being that the positively-loaded 'cost-optimal' wording would lead member states to opt for the realization of more strict and costly regulations.

In a parallel development, MEPs – among them Claude Turmes, known from the renewable energy case – were pressing for building-specific requirements in the renewable energy directive. The renewable energy industry actively supported these moves. As a result, the 2009 Renewable Energy Directive came to state: 'Member States shall introduce in their building regulations and codes appropriate measures in order to increase the share of all kinds of energy from renewable sources in the building sector' (European Parliament and Council 2009a, Art 3.4).

The Commission's final EPBD recast draft was issued in November 2008, as part of a wide-ranging package of measures on energy efficiency (including directives on energy labelling and labelling of tyres) (European Commission 2008h). The proposal signalled an increased focus on renewable energy sources, a mandatory EU-wide formula for calculating 'cost optimal energy performance' levels and the scrapping of the 1000m² threshold in the building code requirements for renovation of buildings. The market flair of the energy performance certificates measure was strengthened somewhat, in that such certificates would have to be used in all sales and rental advertising.

What role did the three groups of construction industry actors play this time? Interviewed EU officials reported that the constructors, represented by FIEC, gave inputs when they were invited to, but did not play a proactive role. The building product producers seem to have been both more active and more professional than ten years earlier. Eurima and EuroACE actively lobbied the Commission and were active in the public debate, repeatedly supporting stronger centralization, a binding requirement for all new buildings to be low- or zero-energy by 2020, and fiscal incentives (*ENDS* 2009a). The insulation actors ordered several reports from technical research institutions, which they used to strengthen their claims. Eventually, also the glass producers argued for stronger centralization, but they were not as active as the insulation actors. Industry interviewees told us that they targeted policymakers based in Brussels, and not national policymakers.

The third group of actors, the building managers, had become slightly more positive to an EU energy policy for buildings. For instance, our interviewees noted that both European Property Foundation and Housing Europe now took a positive stance on energy certification. These two organizations seem to have become more active as a result of the national implementation of the first EPBD (European Parliament and Council 2002). According to EU officials, they still kept a fairly low profile in the deliberations, however.

Interviews indicated that three groups of industry actors collaborated more this time around than in the processes that led up to the first EPBD. Yet, the overall picture is that the insulation industry was the only building construction industry actors that devoted considerable time and energy to the Brussels game. In addition, the renewable energy industry was more involved now (see for instance EREC 2009). According to our interviewees, EUFORES, the pro-renewables MEP organization with strong links to the renewable energy industry, aimed for one of their appointees to get this dossier. It came as a surprise to all when the rapporteur post was given to Silvia-Adriana Ţicău, a young and inexperienced MEP from Romania. However, she soon joined EUFORES after having been appointed rapporteur.

Already in its first discussion in the directive, the European Parliament called for more centralization than proposed by the Commission. Ţicău called for faster, deeper harmonization across Europe, in both energy performance certificates and cost-optimal minimum efficiency standards (*ENDS* 2009c). She also urged stronger fiscal language, not the least because the new member states would need fiscal support from EU funds in order to implement stricter requirements (*ENDS* 2009b). Other MEPs pushed for other amendments that would centralize the power within the issue-area, including requiring all buildings to be zero-energy by 2020 (*ENDS* 2009b). Furthermore, the creation of an EU energy-efficiency fund by 2014 was called for, to finance efficiency improvements in buildings. A new annex to the directive was proposed, listing financing instruments that EU countries should use to promote energy efficiency (*ENDS* 2009d).

In the end, all the above-mentioned amendments were approved by the Parliament with an overwhelming majority (*ENDS* 2009e). According to our interviewees, MEPs were keen to use the EPBD legislation processes in their 2009 election campaigns, especially now that climate issues were high on the agenda in the run-up to the Copenhagen summit. Hence they showed unusual high involvement and engaged in heated discussions. Financing and zero-energy buildings took on high symbolic importance.

Interviewees agreed that it was due primarily to the efforts of rapporteur Ţicău that the Parliament came to press so hard concerning financing (see also European Parliament 2009a). Concerning zero-energy buildings, this was promoted by many MEPs, especially those from EUFORES and the Green parties. Apparently, the zero-energy concept became important in itself, even though it was clear that actually achieving zero energy would be very hard, if not impossible. As one interviewee remarked, 'I must admit that it was a bit annoying that the word was more important than the actual content.'

The immediate response of many member states to Commission's recast draft was rather negative (*ENDS* 2009c). Most agreed with the objectives of the EPBD, and few questioned the holistic method that encompassed both renewable energy and the thermal qualities of buildings. But proposals that would enhance the centralization of power evoked considerable opposition. Since the Parliament's proposals would act to strengthen the centralization of power, they gave rise to strong protests. According to a report from an Energy Council meeting in May 2009, the Parliament's amendments 'appear at first sight to be overly ambitious and unrealistic; the Commission's opinion on these amendments is therefore eagerly awaited' (European Council 2009b: 3). In particular, there was widespread scepticism to focusing more on existing buildings as regards changes in the certificate system and the 1000 m² threshold. After several discussion rounds, member-state representatives reported in April that the recast was proving not to be 'an easy file' (*ENDS* 2009d).

How did the three groups of member states act? According to our interviewees, Denmark was the only really enthusiastic country, although Sweden and Austria were also positive. EU official interviewees noted that Germany was split on the issue, with people from the environment ministry being very positive to higher ambitions and more centralization and the industry ministry representatives far more reluctant. The UK found itself in a special situation, with its national government led by Gordon Brown facing severe political problems at home. One interviewee characterized the situation like this: 'Due to the difficult political climate hardly anything gets done; they are hardly able to make any decisions'. Hence two of the dominant frontrunner countries, Germany and the UK, were hindered from playing a strong role in the Council decision-making processes. As to Spain and other southern countries, they do not seem to have played any major role in the deliberations.

The East European countries seem to have represented the greatest challenge. Our interviewees indicated that this group was negative to all proposals that targeted the existing building stock. Particularly problematic were the certification requirements, which one ·interviewee described as 'the Achilles heel'. Other interviewees noted that, thanks to the 2002 EPBD, many of these countries had, for the first time ever, started to pay attention to the energy performance of their building stock. This may have made them more willing to accept new regulations than they would otherwise have been, but still they were fairly sceptical.

Despite the bleak prospects, a shift in the negotiation atmosphere occurred towards the end of the summer 2009, apparently related to the approaching climate summit in Copenhagen. Because the measures in the EU climate package had been settled already in late 2008, the EPBD recast represented the most significant EU climate policy not yet decided on. Sweden, which had the presidency during the finalizing phase of the EPBD negotiations, was highly aware of the symbolic potential of delivering an ambitious directive and decided to give high priority to the energy-efficiency package (European Council 2009b). The symbolic importance of the EPBD became even stronger because it was scheduled to be decided at a meeting of energy ministers to be held on the opening day of the Copenhagen summit itself.

Sweden pushed hard for a rapid process and announced that it would carry out trialogue negotiations with the Parliament long before the member states had reached agreement (*ENDS* 2009f). Interviewees stated that this trialogue process served to foster greater confidence between the Council, the Parliament and the Commission on this issue. The Commission was now far more ambitious than indicated in its initial draft and supported the Parliament's focus on zero energy as well as financing. However, it proved hard to get concessions from the member states on this latter point, although their reluctance towards granting formal authority to the EU faded somewhat as the Copenhagen meeting drew closer (*ENDS* 2009g, 2009h). According to EU officials we interviewed, no bargaining deals were struck about the three directives in the energy-efficiency package. A Council report also shows that deliberations on the two additional directives proceeded smoothly (European Council 2009b).

It came as a surprise to many observers when a final deal was struck in mid-November (*ENDS* 2009i). The core features were in line with the original proposal from the Commission, although some significant changes had been made. The Parliament had managed to get the national governments to accept 2018 as zero-energy deadline for the public sector, with 2020 as a deadline for new buildings in general. However, the Parliament's 'zero-energy building' wording was replaced by the far more ambiguous 'nearly zero energy buildings' (European Parliament and Council 2010). The agreement said little about financing: member states were merely required to report on their activities in this respect. Despite the local loading character of the outcome, the Swedish Presidency declared that the agreement sent a strong signal to the climate talks that the EU 'can move from words to action on climate' (*ENDS* 2009i).

To a large extent, the actual consequences of the Directive will hinge on how the two new concepts, 'cost-optimal' and 'nearly zero energy', are understood and implemented. Already one year after adoption of the new EPBD it had become evident that differences in interpretations might create challenges. For instance one interviewee stated, 'my worry is the existing building stock. Here people are trying to read it differently from what we intended.'

We conclude that coverage of the EPBD increased significantly as a result of the 2009 recast, but that on the whole, the new version of the Directive involves only incremental changes. To what extent and how have industry, interaction and the external environment influenced policymaking in this issue?

Why has the EU Developed a Local Loading Energy Policy for Buildings?

Industry: Decentralized Industry Underpins Decentralized EU Policy

The low level of industry engagement stands out as a striking feature of the policy process. The EU's energy policy for buildings primarily targets the building construction industry. This complex industry consists of three groups: constructors, building product producers and building managers. While the building product industry is fairly concentrated and dominated by a few big corporations, the two other categories are more fragmented.

Our empirical material shows that from the time this issue-area emerged in the late 1970s and until 2010, most industry actors had no clear positions. Not until 2009 did a significant number of them start to engage in policymaking. The insulation industry is the notable exception: ever since the early 1990s they have engaged strongly in the development of EU policy. First and foremost, their focus has been on promoting measures for technology development, but they have also supported the more market-based certificate scheme launched by the Commission. Hence, they have had a somewhat pragmatic position as regards the steering method. Eventually also the renewable energy actors engaged in this policy area, particularly after the EPBD decisions in 2001 which sidelined renewable energy and energy efficiency as core concerns in EU energy policy for buildings. As discussed in Chapter 5, the renewable energy industry is characterized by increasing centralization and the logic of technology development.

We begin by searching for evidence of the instrumental governments' mechanism. If this mechanism has been employed, we should find a correlation between industry and member-state positions – but in order to maintain that it has been in operation we also need to know more about the processes of national preference formation.

First, and using a long time-perspective, from the early 1970s and up to the present, we see that member-state positions on the competence distribution and steering method appear to have remained fairly stable, although we also note a growing willingness to create an EU energy policy for buildings that would go beyond symbolic statements. Since few industry actors had clear positions at an early stage, they can hardly have instructed or influenced the national governments.

True, the building construction industry has shown a marginal increase in its support for an EU policy, but this is more likely to be the result of greater political attention to the issue nationally and in the EU than the other way round. Furthermore, as it seems to be in the economic interest of most building construction actors for the EU to have an ambitious policy for buildings, we could have expected much stronger member-state support for this kind of policy development if their positions had reflected economic industry interest in a simple way.

Second, we ask whether the positions of the five member states in focus in this book have been shaped by pressures from industry. Due to the national character of most building construction industry actors, all five countries have substantive national industries here. The sheer complexity of the industry makes it challenging to trace the relationship between the national industry actors and the national political positions. Our assessment relies primarily on information from euro-federations and EU officials; we have not systematically mapped the preferences of all industry actors at the national level. This empirical material provides little indication that industry actors have actively contributed in shaping national positions.

Germany has developed a policy with a strong technology-development flavour, in conjunction with EU policy developments. This seems to result mainly from the national political shift in German energy policy, driven more by changing political sentiments than pressure from industry (see Chapter 5 for details). Later than Germany, the UK shifted its position radically, but we do not see any indication that this was industry-driven. Moreover, the fact that UK had implemented a rather radical interpretation of the first EPBD did not mean that it became more positive to greater centralization of authority in this issue-area.

What then of Sweden? It has been a frontrunner in energy efficiency for some three decades, before any industry actors became involved in this policy area. Sweden skilfully used its position as holder of the EU Presidency to ensure that the recast EPBD was decided in 2009, but we find nothing to indicate that this was the result of pressure from industry. And finally, Spain and Poland were reluctant to the development of an EU policy here. This may relate to lack of involvement on the part of their national industries, but it is hard to draw clear analytical conclusions on the basis of such 'non-events'.

We have paid particular attention to Denmark, since it has played a special role in this policy area. Denmark is the country that has most keenly promoted centralization of this policy area and is also the home of prominent insulation producers. Yet, the position of Denmark seems to have been motivated not least by its belief in energy efficiency as a climate-policy measure, and not only by the economic interests of its industry.

In general, the member states have referred more to national regulation traditions and the quality of their building infrastructure than to national industry interests. Moreover, to the extent that industry has affected policy development it seems to have approached other participants in the policy process than national governments. For instance, the insulation industry lobbied the Commission in order to ensure that the recast process would be undertaken, and most of the renewable energy content in the EPBD has resulted from Commission or Parliament initiatives, not proposals from the member states.

This leads us to conclude that the instrumental governments mechanism has played a minor role in this policy area. Due to the complex and decentralized character of the building construction industry, it may be not so surprising that the influence of industry is not easily captured by such a generic theory as Liberal Intergovernmentalism. The member states as such have been central, not least in hindering stronger centralization – but LI falls short when it comes to explaining how the member states have developed their positions.

Moving on to search for institutional feedback mechanisms, we should begin by trying to identify and describe the organizational fields – industry, various commercial organizations, public regulators and EU organizations – that have participated in policymaking. However, we do not find a clearly defined European organizational field for building construction. First of all, the industry is fragmented and characterized by few horizontal links between the three groups of actors (constructors, building product producers, building managers).

The relationship between industry actors within the three groups is also weak – for instance, the various construction professions tend to focus only on their one core activity, be it roofing, carpeting, plumbing, and have little contact with each other. Nor have we been able to trace many strong links between industry actors and EU organizations. The insulation industry and EuroACE seem to be the only actors to have had much contact with Commission officials. Nor does the industry have particularly strong connections to any specific governments. Apparently, then, the situation is that since all member states have a construction industry, no member state takes any particular interest in defending it.

Moreover, the culture of the building construction industry varies significantly from country to country. We have not been able to trace any clear institutional logic that characterizes this field. Although members of the building construction industry are commercial actors, they have not promoted market measures. The initiative for a market instrument – the certification measure – came from within the Commission, not from the industry. Moreover, technical research communities have been the main cradle for policy ideas, not industry actors. The insulation industry has generally promoted the policy ideas produced by technical researchers, not developed ideas itself. All in all, we have seen little conflict between institutional logics during the policy deliberation, as most actors here have been either pragmatic or have supported a technology-development approach.

As discussed in Chapter 5, the organizational field of renewables emerged in the late 1990s and became significantly strengthened during the following decade. We have also noted that the renewable energy industry is embedded in a technology-development logic. It would appear that the renewable energy industry has increased its influence in line with the emergence of a united European organizational field of renewables, which played primarily into the last phase of policy development. This may help to explain why EU energy policy for buildings in 2001 changed from being an energy-efficiency policy to becoming an issue of energy performance, encompassing renewable energy as well, and why the focus on renewable energy got strengthened in the EPBD recast.

The many national organizational fields of building construction, and the absence of a European field, have created specific institutional feedback. First, the Commission held on to their own ideas, in the sense that all major components in their EPBD draft in 2009 can be found in the 1984 Commission proposal that was turned down by member states. The insulation industry has not had the resources needed to develop new policy ideas, but has instead aligned its proposals to Commission initiatives and inputs from technology researchers. Second, due to the absence of industry presence on the European scene, the Commission and Parliament actors have not had actors to play ball with. The insulation industry and EuroACE represent only a small fraction of the industry, so their policy impact has had limited credibility. Lack of industry support has made it hard for the Brussels-based actors to counter member-state inclinations to resist centralization.

Moreover, in order to foster member-state support for EU involvement in this policy area, the Commission has proposed measures that fit with existing practices among the member states. During deliberations on the EPBD, member states have ensured that proposals from the Commission and the Parliament have been even further adjusted to their own national practices. This has resulted in institutional feedback that reproduces existing national steering methods, and prolongs the decentralized competence distribution. This mechanism contributes to explain the low level of centralization as well as the technology-development character of the outcome.

On this backdrop, we may conclude that the fragmented and national character of the building construction industry has obstructed the emergence of a more stable European organizational field around this industry. This situation prevents industry actors from having much direct impact on EU policy processes. The absence of a European field has a profound effect on the EU policy process. First, it gives considerable room to the experts: the policy outcome has been developed mainly in dialogue between expert staff in the Commission and national governments, both drawing on inputs from technical research communities. (Here we may note that the Commission has held on to its initial ideas from 1984, changing these only marginally.) Second, it reduces the clout of the few industry actors that have actually engaged in the EU policy processes. The insulation industry has been rather heavily involved, but without managing to bring about much harmonization and centralization of competence in this area.

This leads us to conclude that institutional feedback has set its mark on the final policy outcome. Yet, it cannot provide the full explanation as to why this EU policy emerged in the first place, or why it should have such a distinct technology-development character.

Moving on to the network entrepreneurship mechanism, we see that the level of entrepreneurial activity has been quite low. Yet, entrepreneurs have involved themselves in order to affect the outcomes on some special occasions. First, the Commission exercised daring entrepreneurship in the 1980s and early 1990s. In

1984, the Commission launched ambitious proposals, but these gained meagre member-state support. Efforts were repeated in the early 1990s, and that led to the adoption of the SAVE Directive in 1991. The Commission has aimed for a fairly high level of centralization and the introduction of a clear market instrument (certificates) – but the final directive proved to be mostly symbolic, with no clear governance character. Hence, the entrepreneurship failed initially.

We do not see the same kind of extraordinary Commission engagement after the turn of the millennium. Due to the institutional feedback effects discussed earlier, the Commission was, however, able reap some benefits from its earlier entrepreneurial efforts. The EPBD did not have the centralized character that the Commission aimed for initially, and the steering method was mainly technical – but it did turn energy policy for buildings into a substantive EU policy.

Second, the insulation industry performed entrepreneurship from 2000 and onwards, and this appears to have helped make the EPBD more substantive. Moreover, pressure from insulation actors may well have contributed to the launch of the recast EPBD in 2009 – although here we should recall that it enjoyed the support of important Commission officials as well. The insulation industry also contributed to develop the 'cost-optimal' language of the final text of the recast EPBD, but this did not change the competence distribution in the issue-area, although that had been the intention. The insulation industry created some network ties to the Commission and the Parliament, but it had a hard time getting other industry actors and national governments to join in. On the whole, then, the insulation industry was not particularly successful, and their entrepreneurship yielded small results.

It can be argued that also the Parliament exercised entrepreneurship in the processes of revising the EPBD in 2009, aiming to create a more centralized policy outcome. Their pressure resulted in two new elements in the policy, on zero-energy buildings and on financing. These are policy components that cannot be traced back to the 1991 SAVE Directive or earlier Commission initiatives. Parliament Rapporteur Silvia-Adriana Ţicău acted creatively and consistently to promote financing issues, but also other MEPs invested significant efforts in this policy area, in particular with respect to zero-energy buildings. The MEPs employed their formal powers in the co-decision procedure to the full. All the same, they were not able to make the outcome more centralized.

What then can we conclude as regards the influence of industry? It seems clear that industry actors have not had much impact on the policy outcome, whether through institutional feedback or through entrepreneurship. The policy outcome has been shaped primarily by the initial entrepreneurship of the Commission and later institutional feedback, reflecting dominant national practices and developments in European research communities. The lack of a European organizational field for building construction has given the experts more room to dominate this issue-area. All the same, the experts have not been able to shift much political authority from the national to the European level: a decentralized industry underpinned a decentralized policy.

Some Interaction with Renewables

The energy performance of buildings has been on the EU agenda for many decades, with energy policy for buildings becoming more strongly linked to the EU renewable energy policy over time. Moreover, both SAVE and the initial EU renewable energy policy (ALTENER) were primarily shaped by a technology development logic. But, in contrast to ALTENER, the SAVE Directive also included a market instrument: the energy certification of buildings.

The directive on energy performance of buildings, EPBD, was developed in the period between 1999 and 2002, at about the same time as the initial EU emissions trading and renewable energy directives. Whereas discussions on renewable energy were influenced by ETS-like market thinking, this was not the case with the EPBD. Some of our interviewees indicated that DG Environment was sceptical to a recast of the original EPBD because that might obstruct the functioning of the price mechanism in the ETS. But eventually, DG Environment emerged as supporters of a stronger EPBD, and there was no continued questioning of the relationship to the ETS.

A probable explanation is that the energy use of buildings is not targeted by the ETS, so there is no immediate conflict between the technology standards and regulations in the EPBD and the ETS (although main building component factors and industries such as cement and glass are included in the ETS). Hence, the slight market element of EPBD seems to have emanated more from the general market hype in the EU in early 1990s, and was not related to the later introduction of market thinking in EU carbon regulation.

We have shown that the shift to a holistic perspective on the energy performance of buildings was related to a general shift in thinking among building construction experts and researchers in Europe. However, it may also have been influenced by the parallel development of the EU policy on renewables. Note here that the indicative renewable energy targets adopted in 2001 were based on energy *consumption* and not production – which meant that energy efficiency could become a possible route for countries to achieve their energy targets.

Turning then to interaction mechanisms, do we find that the EU renewable energy policy affected the character of the EPBD? As to functional interaction, we note that the development of the EU renewable energy policy contributed to strengthen the position of the renewable energy industry, and that this in turn increased the focus on renewable energy in the policy process surrounding EU energy policy for buildings. It is also reasonable to assume that the renewable energy targets provided member states with greater motivation to focus on energy efficiency. On the other hand, Germany did not make active use of its strong position in renewable energy to influence the outcome of energy policy for buildings. Neither did Germany – or any other member state that supported centralization of the renewable energy policy – argue that centralization in this area necessitated a similar development in the EPBD. However, what was involved here was probably a more subtle functional interaction: the repeated references to the energy performance of buildings in the renewable energy directive served to strengthen actors who argued for a greater focus on renewable energy in the EPBD.

What then of entrepreneurial interaction? Here the lenses of bargained interaction are not very helpful: the directives on energy policy for buildings have been developed more as single policies, not in packages alongside other climate policies. True, the 2009 directive was part of a larger energy efficiency package, but there are no striking instances of horse-trading vis a vis other parts of that package.

When it comes to institutional interaction, the EU's renewable energy policy has always been characterized by a technology-development approach like that of energy policy for buildings. This may have contributed to strengthen the technology-development steering method of the revised EPBD – but it could just as well have resulted from the fact that both policies were developed by DG Energy, with its preference for a technology-development logic. Central DG Environment actors (with backgrounds from economics, and responsible for the ETS) have been far more inclined to base their policy proposals on a market logic. Hence, we cannot see that the steering method of the policy on renewable energy has institutionally affected the steering method of the energy policy for buildings.

We can conclude that the energy policy for buildings has developed mainly according to its own internal logic, although functional interaction with the EU renewable energy policy has also been of some relevance. Policy-internal path dependencies and interaction have worked in the same direction, strengthening the technology-development character of the policy.

External Environment: Setting the Pace for Policy Development

Several factors external to the EU have played into EU policymaking in this case. First, the global oil crisis in the early 1970 created the initial incentives for developing energy efficiency policies, but this led mostly to the development of national policies. The EU started to target the energy efficiency of buildings when the Commission launched its ambitious proposal in the mid-1980s. This may be interpreted as a late response to the oil crisis, but the initiative did not immediately lead to any tangible EU policy outcome. Second, important decision situations on energy policy for buildings have correlated with main decisions in the global climate regime three times: in the early 1990s, at the entrance to the 2000s and in 2009. Moreover, the EU has generally stepped up its policy development *before* the crucial decisions are made in the global regime; afterwards, the policy is specified in order to fulfil international obligations.

Thus, there has not been much coercive pressure in this case. It was by no means evident that the soaring global energy prices of the 1970s and then in the first decade of the new millennium would lead to the development of an EU energy policy for buildings. That decision was up to the EU member states. The global climate talks may have created a climate for energy-efficiency initiatives, but the global climate regime lacks specific norms and regulations for the energy performance of buildings. Hence, it is not likely that any influence occurred because of coercive pressure.

Is, then, this relationship an example of institutional diffusion? Also this seems unlikely, since the energy performance of buildings has been little discussed in the international climate regime, and the regime gives more weight to cost-efficiency and market mechanisms – not the steering of technology development. The US development of energy labelling may have served as inspiration for the idea of energy certification, but none of our interviewees mentioned this point, so there seems no reason to see this as a major factor behind the introduction of energy certification. Moreover, most of the EPBD outcome in 2001 had a technology-development character. Hence, institutional diffusion does not stand out as an important mechanism in this case.

Rather, it seems as if the climate regime primarily affected the *pace* of decision-making by providing good conditions for entrepreneurship. Commission officials consistently cited global climate developments to put energy efficiency on the EU agenda, each time managing to push the policy result a little closer to their original proposals. The initial Commission proposal – launched in 1984 – was not related to the climate issue at all. It may be that since the Commission and energy-efficiency promoters were able to link the issues of climate change with energy efficiency already in the early 1990s, this led to some kind of automatic coupling of the two issues. However, this link cannot have been particularly strong. At every stage in policy development, some actors have argued that critical stages in global climate regime developments should be linked to buildings-related policy decisions.

Also the EU member states have used the global climate regime as an internal bargaining chip. This is particularly evident in the final phase of policy development, when Sweden played the global card to get reluctant member states to accept the concessions needed in order to reach agreement. Sweden had the EU Presidency in the run-up to the Copenhagen climate summit, which provided an extraordinary chance to exercise international institutional entrepreneurship. Here we note that Sweden did not use international institutional entrepreneurship in order to change the steering method in this policy area, but to strengthen the technology-development element and ensure that agreement on the policy would be reached in time. Also actors in the insulation industry made active use of global climate arguments – in both policy processes in the 2000s – but focusing more on providing facts and figures to underscore their argument, than changing peoples' perceptions of the implications of the climate regime.

Here we see the same kind of international entrepreneurship as in the case of renewable energy: a strong institutional flair. Entrepreneurs used the global situation to spread the view that carbon mitigation was an urgent concern and to create the understanding that the internal policy actions of EU were vital to the global outcome. This international institutional entrepreneurship played two significant roles in policy development. First, it created windows of opportunity, enabling entrepreneurs to press for gradually more substantial outcomes. Second, it made member states more inclined to accept EU interference in this policy area.

This was the case in the early 1990s with the SAVE Directive, and was repeated in 2001, when the first EPBD was adopted.

In 2009, global climate arguments (from Sweden in particular) led reluctant member states to yield and accept a policy outcome that was significantly more comprehensive than the 2001 outcome. Yet, we should not exaggerate the importance of international entrepreneurship in this case; despite the global framing, the member states remained unwilling to develop a centralized energy policy for buildings.

On this backdrop, we conclude that external developments can help to explain the initiation and pace of policy development, but the technology development steering method and the decentralized outcome of the energy policy for buildings still primarily reflect conditions within the EU itself.

Conclusion: Lack of Industry Participation created EU Policy Vacuum

This chapter has sought answers to two questions: Why has the EU developed an energy policy for buildings? and why does this policy have a local loading character? We have seen that the almost total absence of industry has had severe ramifications for policy development: because no European organizational field exists, member state governments have the upper hand. In the 1980s and 1990s the Commission tried to act as entrepreneur: the policy in 2010 can be seen as a result of these early entrepreneurial efforts. The Commission repeatedly used important decision situations in the international climate regime to put the energy performance on the EU agenda. This explains why the issue appeared as an EU climate policy issue. To some extent, the EU's energy policy for buildings has been affected by interaction with its renewables policy, but this has not brought significant changes in steering method and competence distribution.

The decentralized and fragmented industry structure and lack of a European organizational field of building construction industry have created institutional feedback mechanisms that undermine centralization. Initially, the Commission aimed for a far more centralized outcome – but with no industry presence in Brussels, it was impossible for the Commission to conduct successful network entrepreneurship. Hence, we may conclude that the local loading outcome is a result of failed Commission entrepreneurship. Indeed, we will go as far as to suggest that this case indicates that the existence of a Europeanized organizational field is a precondition for strong centralization of power in an issue-area.

Few building construction actors have had stable preferences with regard to EU policy development, and the member-state governments did not act as instruments for industry. Instead, member-state officials have tended to follow their traditional regulatory approaches. Moreover, the lack of industry engagement created a power vacuum in Brussels, giving other actors greater room for action. Commission officials, technical researchers, national civil servants from frontrunner countries and insulation industry actors – all these have performed entrepreneurship at various junctures.

Commission officials have used developments in the global climate developments to set the pace in EU policy development. This ensured that the issue of energy policy for buildings featured regularly on the EU agenda, and was a precondition for the emergence of a distinct local loading policy. However, the link to the global climate scene was always open to interpretation; entrepreneurs repeatedly had to make this link anew. Since this was not done to any extent after the turn of the millennium, this policy area was not included in the 2008/2009 EU climate package. However, the Swedish Presidency managed to ensure that the run-up to the Copenhagen climate summit in December 2009 nonetheless contributed to produce a significant new policy outcome.

Chapter 8

Comparisons and Conclusions

Introduction

This chapter takes stock and carries out a focused and structured case comparison, with a view to the main question of this book: *How can we best explain the development of EU climate policy?* As we have seen, EU climate policy is far from being a coherent policy area, with sub-issues characterized by different policy characteristics: the ETS is a Single European Market policy, Renewables and CCS can be characterized as EU Engineering policies, and the energy policy for buildings can best be characterized as Local Loading policy.

As noted, earlier research has indicated that industry, policy interaction and the environment external to the EU are all important for EU climate policymaking, but the paths or features that link together causes and effects have not been well understood. In other words: we need to explore the mechanisms at work. Case comparison allows us to specify the mechanisms that drive EU climate policymaking. Throughout this study, we have used three dominant European integration theory approaches as central analytical signposts – Liberal Intergovernmentalism, New Institutionalism and Multi-level Governance – from which we have extracted specific mechanisms. In this chapter we offer some conclusions as to which of the mechanisms that actually drive the development of EU climate policy, and discuss possible systematic relationships between industry, policy interaction and external environment conditions and policy outcomes. To indicate what will be discussed in this chapter:

- First, industry was not a mere braking block, but rather a quite multi-faceted actor. Moreover, the Europeanization of industry and the character of the organizational fields in which industry takes part have important implications for how industry will influence EU policy. This has implications for future EU climate policy research, but also relevance for general European integration theory.
- Second, strikingly little mingling and cross-fertilization took place between the four policies in focus in this book. We discuss why this has been the situation in EU climate policy thus far, and why it may change in the future.
- Third, the external environment has primarily been important in a different way than usually highlighted: we find that policies and ideas stemming from international regimes have been creatively used by entrepreneurs, notably in the Commission, to create and exploit windows of opportunity. We discuss this finding in light of various views on the Commission's ability to exert entrepreneurial influence on EU policymaking.

- Finally, a striking finding is that several different kinds of entrepreneurship have been at work. The Commission conducted a slow and steady 'tortoise' type of entrepreneurship, often related to the EU-external environment. Other Brussels actors seized the opportunity and exerted *carpe diem* entrepreneurship as climate-policy windows were opened. We discuss the general relevance of this.

Policy Outcomes: Centralization and Technology Development Steering Dominate

We have focused on how four selected EU policies – the emissions trading system (ETS), renewables, carbon capture and storage (CCS), and energy policy for buildings – can be located along two main dimensions in EU integration theory: steering method and competence distribution. The former dimension pertains to the type of measures that are applied, while the latter refers to the distribution of authority between national governments and the EU.

We distinguish between two principal steering methods: technology development and market approaches. Technology development approaches generally involve public regulations aimed at promoting certain technological practices or fostering specific industry activities. Market approaches generally entail the creation of new markets or altering existing ones; such approaches do not favour any specific industries or technologies, but tend to rely on market forces to choose winner industries and technologies. By combining these two dimensions, and based on the case studies, we can place the policy outcomes as shown in Table 8.1.

Table 8.1   Scores of the focused policies

Competence distribution	Steering method	
	Technological development	*Market*
Decentralized: Many national fields	1) Local loading Energy policy for buildings	2) Piecemeal market Initial emissions trading system (ETS 2003)
Centralized: Dominated by one Europeanized field	3) EU engineering Renewable energy policy Carbon capture and storage (CCS)	4) Single european market Revised emissions trading system (ETS 2008)

We have placed *EU energy policy for buildings* in square 1, 'local loading'. The Energy Performance of Buildings Directive (EPBD) requires member states to adopt minimum energy requirements in their national building codes, as well as systems for regular inspection of heating and air-conditioning systems and national action plans for near-zero-energy buildings measures – all of which rely on technical requirements determined by the national governments. Thus, all these components have a local loading character. The EPBD also includes an energy certification of buildings component, but it is up to the national governments to decide whether this is to be designed as a market instrument, pitched to affect pricing mechanisms of national markets for buildings, or as regulatory measure that promotes certain technological solutions.

Renewable energy policy is placed in square 3, 'EU engineering'. The 2009 EU Renewables Directive sets a Community-wide target of 20 per cent renewable share of gross final consumption of energy by 2020. All member states are allotted binding individual targets for renewable energy, to be achieved by 2020, and they are required to develop national action plans in line with a detailed template. We see that the EU enjoys considerable authority in this issue-area. The 2009 Directive establishes three flexible mechanisms, allowing member states to collaborate in achieving the goals, but member states are free to decide whether to use these or not. The Directive does not directly require the use of feed-in schemes, but its many technical requirements underscore the importance of developing technologies irrespective of the cost. In this the Directive seems more aligned to feed-in schemes than to market schemes.

Also the EU's *CCS policy* can be seen as 'EU engineering'. As regards the steering method, policymaking here basically adheres to a technology-development approach whereby policymakers finance R&D in order to make technologies more mature and contribute to their dissemination. Even though the level of funding depends on the sale of ETS allowances (the NER 300 fund), how this money is used will be determined primarily by technological criteria, not economic criteria or market measures. As to the distribution of competences, the contribution to EU CCS pilot projects from EU infrastructure funds, the NER 300 and the involvement of the European Investment Bank all indicate significant central steering.

The *revised EU emissions trading system* (ETS) formally adopted in 2009 goes a long way towards establishing a Single European Market for carbon emissions, so we have placed the ETS in square 4. This revised system, to be put into operation in 2013, involves a single, EU-wide emission cap, based on the target of a 21 per cent reduction by 2020 in relation to 2005 levels. National allocations will be derived from this common cap. The main principle for allocation of allowances is auctioning. This will be phased in gradually, with initially far more auctioning of allowances for energy producers than for energy-intensive industries. In addition, the process of handing out free allowances is further harmonized, based on common state-of-the-art technology benchmarks. The Commission is given a prominent role as main overseer and designer of the further development of the system. Overall, the ETS after 2013 will be far more market-streamlined than any other of the three policy areas, and the degree of centralized control will also be strong.

Thus, we see that the policy outcomes have varied greatly. While all policy areas started out as more or less decentralized, three areas have ended up with fairly centralized competence distribution (the ETS, renewables and CCS), while the energy policy for buildings has remained decentralized. This makes it natural to ask: first, *why have we seen a shift towards centralization in the three first areas, but not in the latter?* Second, three of the policy areas have a technology development character (renewables, CCS and buildings), while only one (the ETS) has a market character. *Why does the steering method of the ETS differ from the other areas?*

Let us see which mechanisms can shed light on these similarities and differences, and then examine the related analytical and theoretical implications.

Industry: No Policy Initiator but Multi-faceted Key Actor

Case Comparison: Industry Matters through Institutional Feedback and Entrepreneurship

Case comparison offers a foundation for assessing how much and in what ways the emergence and character of climate policy outcomes have been shaped by the three mechanisms of instrumental governments, institutional feedback and network entrepreneurship:

- *Instrumental Governments:* dominant member states defending the interests of their economically powerful national industries (drawing on Liberal Intergovernmentalism, LI);
- *Institutional Feedback*: the dynamics of the organizational fields involved in policymaking (New Institutionalism, NI);
- *Network Entrepreneurship:* industry actors with exceptional skills or intensity in the policy process (Multi-level Governance, MLG).

Table 8.2   How industry shaped policy outcomes – rough scores

Industry mechanism	Issue area				
	ETS	*Renewables*	*CCS*	*Buildings*	*Overall importance of mechanism*
Instrumental governments	Some	Little	Little	Little	Little
Institutional feedback	Some	Significant	Some	Significant	Significant
Network entrepreneurship	Some	Significant	Some	Little	Some

Table 8.2 summarizes and compares the extent to which the three mechanisms affected the character of the policies. In the following we again present the background for the rough scores, as well as discussing how industry affected the initial emergence of the policies.

We start with *the ETS*. This market measure was certainly not initiated as a response to strong demands from industry. Instead, as have seen, entrepreneurship from the Commission was crucial for the initial EU adoption of this instrument. On the other hand, the development of the initial system was heavily influenced by the dominant industrial group and main target for the system – the utilities. The energy-intensive industries were rather passive at this stage. The 2009 'governance revolution', which led to a far more centralized and market-streamlined scheme, was not initiated by industry either, but in a somewhat similar way as in the initiation phase it was supported by many utilities. And this time, the energy-intensive industries sought to obstruct much of this development.

In neither of the two decision-making processes did the member states act as instruments for their industries. Developments are more aptly caught by the NI mechanism. We see two contrasting institutional feedback effects at work here. First, developments within the European organizational field of electricity production strengthened the Commission's case for more of a purely market design, with greater centralization of control. Somewhat paradoxically, the initial malfunctioning of the scheme led utilities and member-state governments alike to call for improvements of the existing system – with greater market streamlining and centralization – rather than switching to alternative regulatory routes. The Commission had been given a sufficiently prominent position in the first ETS directive to ensure a strong hand in the revision process. This helped the Commission to facilitate the revision process in a way that underpinned the growing support towards an ETS with a Single European Market character. And the debate about windfall profits made it harder for the utilities to maintain their initial reluctance towards auctioning.

Second, we see somewhat the converse when it comes to feedback effects within the organizational field of energy-intensive industries. Experience with the first phase of the ETS led this field to unify in calling for exemptions to market streamlining of the instrument. Interestingly, these field actors did not challenge the existence of the ETS as such. This must be understood as a consequence of the prestige of the ETS as the cornerstone of the EU climate policy: any question of abolishing the system was simply not on the table.

Moreover, the energy-intensive industry actors were not able to launch a credible alternative to emissions trading. The proposed baseline-and-credit design failed to attract widespread interest and support, whether among the utilities or EU officials. On the whole, the feedback effects from the initial ETS within the organizational field of electricity production created greater pressures towards a Single European Market design than the countervailing feedback effects within the field of energy-intensive industries.

Network entrepreneurship has been a central ingredient of this process. The Commission was central in the launching of the ETS and instrumental in setting in motion and shaping the 'revolution' that led to the adoption of a more Single European Market type of system in 2009. Arguably, important reasons for the Commission's success in getting its preferred system adopted in 2009 can be found in clever manoeuvrings a few years previous, creating institutional feedback. As to more specific industry entrepreneurship, the most notable example is the 'campaign' of the energy-intensive industries which resulted in the adoption of significant exemptions from the dominant drive towards a Single European Market system from 2013 on. The industry not only set in motion instrumental government mechanisms in motion, they also operated through multi-level networking.

All in all, operation of industry mechanisms did contribute to the outcome – but it is also clear that industry influence was not the most decisive factor. The electricity producers contributed to underpin the entrepreneurship of the Commission, while the energy-intensive industries managed – to a certain extent – to reduce the effect of the Commission's entrepreneurship. Later in this chapter we discuss how the Commission achievements were conditioned by influences from outside the EU.

Renewable energy attracted attention from both the traditional electricity producers and the renewable energy industry. However, the EU policy on renewables was not initiated by the industry: the EU had started to develop such a policy already in the 1980s. Back then, the electricity industry was uninterested and the renewable energy industry did not exist. From 1997, we see increasing involvement on the part of industry: once the renewable energy industry had achieved a certain size, it called for greater EU engagement. And as the EU and the member states developed increasingly stronger policies the traditional electricity industry in turn stepped up its engagement.

There is a certain match between industry and member state positions; countries with a renewable energy industry favoured an EU engineering policy, whereas countries with no renewables industry preferred market measures or did not adopt a clear position. However, the various national governments did not act as instruments for economically dominant industries. This is particularly clear in the case of Germany: the German government took common positions with the renewable energy industry, not with its electricity industry, even though the latter being far stronger in economic terms. Hence, instrumental governments was not an important mechanism in the case of renewables. Instead, the similarity in the positions of industry and governments seems to stem from common regulatory traditions: over time, both developed common views and preferences with regard to steering methods.

The EU engineering outcome was highly influenced by institutional feedback effects within the organizational field of renewable energy: this industry has long been supported by feed-in mechanisms, and has adjusted its way of operation and business strategies to this type of support. As the renewable companies Europeanized their activity they came to favour stronger EU involvement, but they resisted the introduction of market measures. Such measures like green certificate schemes would only favour the most profitable technologies in renewable energy (such as hydro- and wind power), rather than allowing a whole array of technologies (such as photovoltaic and geothermal heating) to be installed and refined. The strong ties between renewable energy industry actors, German and Spanish governmental officials, Commission officials and European parliamentarians served to make this institutional feedback effect strong.

Also network entrepreneurship played a part. MEPs as well as industry actors showed exceptional skills and intensity in creating multi-level networks and exploiting windows of opportunity. This entrepreneurial activity led more and more countries to support an EU engineering outcome – even the UK, which had originally promoted market solutions, eventually changed its stance. Commission officials and traditional utilities tried to champion market solutions, but without much entrepreneurial success.

We conclude that industry had little influence on the emergence of an EU policy for renewable energy, but that industry mechanisms were highly important in shaping the outcome and actual character of the EU policy. The LI approach – with its instrumental governments mechanism – offers little help in capturing how this industry affected the final policy outcome. It was not economic strength, but the character of the organizational field that made this industry powerful in Brussels: the EU policy was shaped by forceful institutional feedback mechanisms. In addition, the renewables actors succeeded in their entrepreneurial efforts to get powerful member states to support their views.

Turning to the case of Carbon Capture and Storage (CCS), we note that industry was more involved in joining the EU CCS policy bandwagon than in pushing it. The dominant pattern is that of Brussels actors spearheading the idea of an EU-engineering CCS policy; the positions of member states seem to have been influence more by Brussels discussions than by claims and inputs from the their national industries. Arguments relating to the interests of the electricity industry did induce member states to change their positions during the policy development processes, but these arguments were not rooted within industry itself. Instead, they had been 'invented' by entrepreneurs within a Brussels-based CCS policy network – and the industry did not adopt the view that CCS was in its own interests until very late in the policy process. The Dutch and British governments had positions rather similar to those of their oil corporations, but they developed similar interest perceptions in conjunction: this was not a one-directional influence from industry preferences to government positions.

The short history of European CCS policy makes the mechanism of institutional feedback less prominent here, and the only long-term institutional developments were found within the oil industry. In terms of explaining the character of the policy outcome, this perspective mainly sheds light on the steering method chosen. The technology development character of this policy reflects the institutional logic of the experts engaged in this policymaking – mainly people from the research units of the oil industry. These experts were engineers and thus not as marked by the market logic as was the case at the corporate headquarters of this industry.

To some extent, industry actors acted in line with the network entrepreneurship mechanism; they established networks, and sought out and utilized a window of opportunity. The entrepreneurial efforts of MEPs were also noteworthy, with the Commission playing more of an enabling role. The Commission was important for creating the central window of opportunity, whereas MEPs, the CCS-promoting industry and environmental organization actors coordinated their efforts more closely. On the other hand, the entrepreneurial success in this case was partly also due to developments in the outside environment, as discussed later.

In contrast to the cases with a longer history, the factors that shaped the emergence of the EU CCS policy also shaped the character of the policy. Industry affected the policy outcome mainly through entrepreneurship, with institutional feedback playing a lesser but still noticeable role. However, the saliency of industry influence should not be exaggerated; the main explanation for this policy outcome is to be found elsewhere.

With the EU's *energy policy for buildings*, industry actors were definitely less prominent. However, we argue that it is precisely the lack of active industry engagement that held back centralization of the EU policy in this issue-area. This policy – as with all the others – was not initiated by industry. It was the Commission that first championed the development of an EU energy policy for buildings. This resulted in the adoption of a policy in the early 1990s, but the initial policy was only symbolic and of little practical importance.

The EU policy targets a whole array of building construction industry actors, but very few of these developed clear positions initially. The exception was the insulation industry, which began to get involved already in the early 1990s. Yet, since most of the industry actors had no clear positions, we can conclude that the instrumental government mechanism was not prominent. Although most industry actors would probably profit economically from an EU policy in this area they did not call for such a policy to be developed.

In contrast to the other cases, industry was not an important carrier of institutional feedback effects. The multi-faced industry was basically local in nature and lacked wider contact with other EU actors. Hence, a centralized, European organizational field of building construction did not exist. This made it difficult for the Brussels-based actors (like Commission officials, some MEPs and insulation industry representatives) to influence the positions of member states. The member states based their positions mainly on their own regulatory traditions; national governments sought to ensure the continuation of their practices. Hence, the lack of an European organizational field created institutional feedback effects that served to stimulate continued decentralization in this issue-area.

Insulation actors were the only industry representatives to exert any sort of entrepreneurship. However, since they lacked strong ties to other construction industries and national governments, it was not easy for them to create networks or utilize windows of opportunities. Commission officials operated as eager entrepreneurs in the 1980s and 1990s, but failed to achieve much. As with the insulation industry, their entrepreneurial efforts received little response from industry actors and member states. Eventually the Commission stopped expending so much energy and creativity on this issue, although global climate policy cycles have continued to bring it up. The character of EU energy policy for buildings can be seen as very much a result of failed Commission entrepreneurship.

Theory Implications: Organizational Fields Shape Policy Outcomes

The first conclusion to be drawn is that industry has played a multifaceted role, and not merely as an obstructive braking block with 'vested interest in securing the deregulation' of climate policy (see Hix 2005: 69). In line with the proposition of Grant (2011) we conclude that industry was no agenda-setter and was rather irrelevant to the emergence of policy. However, industry structure and activity had high importance for the character of the renewables and the buildings policy outcomes, and some importance for the ETS and CCS. Initially, we expected industry to have been more important for the ETS and renewables than the two other cases. The main reason that our findings depart from our expectations in the case of buildings is that a decentralized and fragmented industry structure created institutional feedbacks that effectively undermined the centralization of EU policy. Interestingly, industry was just as important for the outcome when it lacked resources for lobbying as when it engaged in intense Brussels campaigning.

The industry not taking the initiative may be seen as challenging the LI view of industry as the driving force in EU policy developments. However, it resonates well with the views of MLG scholars and other writers who see the Commission and other Brussels actors as important policy initiators (see Hooghe and Marks 2001: 13, Marks and Hooghe 2004, Hix 2005: 68).

Despite its irrelevance for policy emergence, industry is crucial for the character of the actual policy – but it hardly operates according to LI tenets. The instrumental governments mechanism appears to have been in operation with the ETS case, but was not significant in any of the other cases. The institutional feedback mechanism proved far better suited for capturing how industry influenced the distribution of competences, as was especially evident in the cases of renewables and buildings. As we will discuss in further detail below, those two cases are especially interesting, since institutional feedback underpinned greater centralization in the former while hindering centralization in the latter.

Moreover, the ETS, renewable energy and CCS outcomes all reflect the institutional logic of dominant industry actors: the ETS was underpinned by the market logic that gained hold among utilities (and partly also energy-intensive industries); the renewables policy was shaped by the technology-development logic of the renewable-energy industry; and the CCS outcome was influenced by the technology-development logic of the oil companies' R&D departments. The energy policy for buildings reflects the institutional logic of the various experts (in the Commission, governmental organizations and research groups) involved in the policymaking, not the institutional logic of industry actors.

What was the role of network entrepreneurship when it comes to explaining how industry contributed to shape climate policy outcomes? Table 8.2 shows that its influence has varied from 'significant' (renewables), to 'some' (ETS, CCS), and 'little' (buildings). The ETS and CCS are similar in that both cases were shaped by Commission entrepreneurship; industries joined the Commission bandwagon rather than setting it in motion or pushing it. However, the case of renewable energy showed that industry could perform effective entrepreneurship. Here, a Brussels-based industry contributed to change the preferences of governments. The case of buildings is special, with very little entrepreneurship on the part of industry. Moreover, the Commission spent several decades seeking to promote the same solutions, without fully succeeding.

The generally superior role of the Commission is not surprising when seen in light of MLG arguments. But it implies that scholars like Hix (2005: 230) who have argued that business interests are 'particularly capable of playing the Brussels game', may be overstating the entrepreneurial role of economic interests in Brussels.

Moreover, we note a difference in the entrepreneur's commitment and patience. The Commission operated as a slow and steady 'tortoise' in the ETS and renewables cases, working towards the same objectives for long periods of time (with more success in ETS than in renewables). In the CCS case, it was *carpe diem* entrepreneurship that dominated; after a special window of opportunity for climate policy initiatives opened in the 2005, a whole range of actors started to call for an ambitious CCS policy. Note here that few of these actors had earlier actually promoted CCS. For instance, MEP Chris Davies became involved in CCS only after he had been appointed Parliamentary Rapporteur for the CCS Directive. From that point on, he exerted network entrepreneurship with extraordinary intensity and political skills. It is reasonable to assume that Davies and other CCS proponents wanted to make an imprint on EU climate policy, and since CCS was an issue in which they had a good chance of making a difference, they became involved. As we have seen, the efforts of Davies and industry *carpe diem* entrepreneurs were crucial to the CCS outcome. However, such changing positions seem to run counter to entrepreneurship studies emphasizing that entrepreneurs tend to have rather fixed policy ideas (see for instance Meijerink and Huitema 2010).

Whether these actors have become truly committed to CCS and will continue to promote this policy solution is yet unclear. However, we may note that when the Commission first promoted emissions trading this was also in a sort of *carpe diem* manner: the Commission had not previously been committed to emissions trading (to some extent, quite the converse), but turned to this measure much because the carbon tax proposal had failed. Eventually, this became a long-term tortoise commitment. We return to the tortoise/*carpe diem* dichotomy in our subsequent discussions on interaction and the external environment.

As indicated, our industry findings challenge the relevance of LI theory in explaining how industry affects EU climate policy. Do they also undermine the theory in general – or does this simply mean that Moravcsik (1998: 37) was right when he indicated that EU environmental policy was special, in that politicians tend to have 'independent preference for regulatory goals' in environmental policy? We think that our findings have rather deep ramifications. Industry was important for the policy processes, and governments did indeed depend on industry, but this did not play out in the way described by Moravcsik, as a unilateral development in which industry dictates government positions. What we have seen are multi-directional social processes where both industry and national governmental bodies acted on the basis of the institutional logic and the authority structures in which they were embedded – much in the way described in the NI literature on organizational fields (see DiMaggio and Powell 1991, Scott 2008, Fligstein and Stone Sweet 2002).

Moreover, we claim that the instrumental governments mechanism neglects long-term developments, which were prominent in our cases. Industry, EU organizations and national governments have made commitments based on existing institutions and policies, and historical policy decisions have structured how industry – as well as other actors – develops new policy positions. This is in line with the NI perspective of an institutional feedback mechanism, as developed by Pierson (1996, 2004) and Fligstein (2008).

Case comparison enables us to develop theory conclusions and new propositions as to the relationship between industry characteristics and EU climate policy outcomes. First, we conclude that *the organizational field's authority distribution shapes competence distribution in outcomes.* Climate policies rooted in centralized organizational fields have become increasingly centralized. The Europeanization of industries has tended to spur the creation of EU organizational fields involving governments, industry and various Brussels actors, with positions shaped by common European socialization processes. In contrast, if industries relate primarily to their own national organizational fields, the position-making processes will remain national. This finding is in line with Fligstein's (2008: 215) argument that 'as people become more Europeanized, their preferences for policy coordination might shift to the European level'.

Varying types of institutional feedback, related to differences in the Europeanization of the organizational fields, help to explain why there has been a shift towards centralization in all cases except the EU energy policy for buildings. Still, we hold that, since CCS policy was not firmly rooted in any particular organizational field, it is necessary to search for other explanations for strong policy centralization here. Note however that centralization of the organizational field and the EU policy outcome can develop in conjunction: with regard to the ETS as well as renewables, the organizational fields have undergone a gradual centralization process in tandem with the development of policy.

Our second industry finding is that *the institutional logic that dominates the organizational field infuses the steering method of the policy outcome.* Fields dominated by market logic have tended to produce market-based EU policies, whereas fields dominated by technology development tend to produce technology-development EU policies. Differences in the institutional logics of the organizational fields contribute to explain why three of the policies in focus (renewables, CCS and buildings) have emerged with a technology development character, while only one (the ETS) has a market character. Industry is often the main carrier of the logic of a field, although the case of energy policy for buildings reminds us that this is not always so.

The logic of the field and the steering method of EU policies adjust to each other over time, and shifts in logics at field level can work in conjunction with shifts in methods of policy steering. This is particularly clear with the ETS, where the character of the first ETS created feedback effects that served to make actors within the organizational field of electricity more positive to market streamlining over time.

These two conclusions indicate that the character of the organizational field(s) can help to indicate the kind of climate policy outcome to be expected. Hence, we can answer 'yes' to our initial question on whether there is any clear relationship between the character of the industries policymaking and the four types of climate policy outcomes. Table 8.3 combines our conclusion on the institutional logic steering-method dimension and on the dimension of the distribution of authority-competence.

Table 8.3   Relationship between organizational field characteristics and EU climate policy outcomes

Authority distribution	Dominant institutional logics	
	Technological development	*Market*
Decentralized: Many national fields	1) Local loading Energy policy for buildings	2) Piecemeal market Initial emissions trading system (ETS)
Centralized: Dominated by one Europeanized field	3) EU engineering Renewables policy (Carbon capture and storage (CCS))	4) Single European market Revised emissions trading system (ETS)

Table 8.3 gives rise to four specific suggestions concerning the relationship between organizational field characteristics and climate policy outcomes:

- Local Loading climate policy tends to develop when the policy process is embedded in decentralized organizational fields dominated by technology-development logic
- Piecemeal Markets climate policy tends to develop when the policy process is embedded in decentralized fields dominated by market logic
- EU Engineering policy tends to develop when the policy process is embedded

in centralized field(s) dominated by technology-development logic
* Single European Market policy tends to develop when the policy process is embedded in centralized field(s) dominated by market logic.

Table 8.3 indicates that MLG scholars may be exaggerating when they argue that 'European states are losing their grip on mediation of domestic interest representation' (Marks, Hooghe and Blank 1996: 341). While this may hold true in some issue-areas, it will be less so for others. We argue that preference formation patterns – as regards steering method as well as competence distribution – tend to be bottom-up in decentralized fields and top-down in centralized fields. In decentralized fields, member-state positions are shaped by national factors. For instance, historical traditions in the building sector shaped member-state positions in the case of EU energy policy for buildings. This situation is rather similar to Moravcsik's view, even though industry has not been such a central actor. With the ETS and with policy on renewable energy, by contrast, preference formation grew less dependent on national traditions as the organizational fields became Europeanized, more in line with the preference formation described by the NI approach and neo-functionalism.

Table 8.3 also indicates a 'chicken/ egg dilemma': does the shift towards centralization and towards market steering *start* with EU policy initiatives or industry-driven changes at the organizational field level? Our research has shown the mutual dynamics at play between EU policy decisions and industry initiatives, making it impossible for us to offer a general and clear answer to the question of which came first.

We also note that differences in industry entrepreneurship tend to co-vary with organizational field characteristics. Thus, our third conclusion concerning industry is that *we see more industry entrepreneurship in European organizational fields than in decentralized organizational fields*. Deeper comparison of renewables and buildings can be useful here. As we have seen, the character of the organizational fields produced opposite effects – centralization for renewables and decentralization for buildings – with industry entrepreneurship playing an important role in the former and an unimportant role in the latter. Thanks to the Europeanized field, the renewable energy industry had a good starting point for entrepreneurship. As the industry was rather well connected at the outset, it was not particularly difficult to approach new actors outside the field, such as government officials and MEPs. By contrast, in the case of energy policy for buildings, neither the insulation industry nor the Commission had network ties to other actors involved in building construction. This meant that they lacked legitimacy to speak on behalf of the whole industry and the resources needed to create network ties to new actors. Thus we see that there is a complex interdependency between institutional feedback and entrepreneurship.

Policy Interaction: Slight Push for Centralization So Far, but Landslides Ahead?

Case Comparison: Primarily Bargained Interaction

In the following, we assess to what extent and how the emergence and character of climate policy outcomes have been shaped by three types of interaction.

- *Functional Interaction:* functional pressure from related policies (LI)
- *Bargained Interaction:* actors that initiate and bargain linkages to other policy area (LI and MLG);
- *Institutional Interaction:* the character of historically dominant policy areas (LI).

Table 8.4   How interaction shaped policy outcomes – rough scores

Interaction mechanism	Issue-area				
	ETS	*Renewables*	*CCS*	*Buildings*	*Overall importance of mechanism*
Functional	Little	Little	Little/some	Little/some	Little
Bargained	Little	Little	Significant	Little	Some to little
Institutional	Little	Little	Little	Little	Little
Persuasion	Little	Some	Little	Little	Little

Table 8.4 summarizes and compares the importance of mechanisms for the character of the four policies under study. In addition to the three mechanisms presented in Chapter 2, it includes *persuasion interaction,* a fourth mechanism detected in the renewable energy case. We will now give a brief presentation of the cases, discussing how their emergence and character have been affected by policy interaction.

Starting out with *the ETS:* was this case affected by functional interaction? The decentralized ETS established in 2003 was still more centralized than the other climate policy outcomes at the time, and was the only issue-area characterized by market thinking. At the time, the recent EU renewables and energy-efficiency policies had not been in operation long enough to bring about functional consequences that could feed into the ETS policy process. However, taking energy market liberalization into account, it is certainly possible that the mergers and acquisitions and the emergence of bigger utilities helped to change the organizational field of electricity production, gradually facilitating the process of developing a far more centralized ETS. But here it should also be noted that energy-market liberalization was only partly driven by EU policy. All in all, the ETS policy outcome was not much influenced by functional interaction.

We have seen that the ETS processes were considerably influenced by entrepreneurship, so perhaps entrepreneurs also applied bargained interaction in order to increase their influence? We do find some relevant bargaining trade-offs in the final climate package negotiations, where countries managed to water down the revised ETS by agreeing to stricter policies in other issue-areas. This contributes to explain why certain exemptions to the European Single Market were introduced, but it cannot shed light on the 'revolutionary' character of the ETS revision as such.

Turning to institutional interaction, there was some important overriding such interaction at play, with the ETS more diffusely inspired by the market thinking so central to the EU as a social project. On the other hand, the ETS differs from other EU climate policy cases that have been characterized mainly by technology-development logics. Moreover, our empirical basis does not indicate that institutional thinking from other policy areas provided a direct influence on the ETS policy process. On the whole, we would say that institutional interaction was of little importance. On this backdrop, our main conclusions are that the emergence of the ETS was not affected by interaction, and that such mechanisms have affected the character of the final policy outcome only very moderately.

Moving to *renewables*, we would expect this case to have been functionally influenced by the EU energy-efficiency policy since the two were conceived at about the same time. However, policies in both issue-areas were initially mainly symbolic, thereby producing few tangible consequences – also for each other. One might also expect that once the ETS had created a carbon price, there would be less need for a strong EU policy on renewable energy. However, since the ETS did not in fact lead to a stable and high carbon price, no functional interaction took place in renewables development. To some extent the ETS and the EU's policy on renewable energy target the same main industrial actors, the electricity utilities in particular. But whereas this group achieved a prominent position in the ETS context, it had less success in relation to renewables. The utilities simply did not manage to transfer political standing and clout from the one context to the other.

Do we find any instances of bargained interaction? The answer is clearly no. We have not seen any actors actively linking EU energy-efficiency policy to the development of a policy on renewable energy during the first decades of policy development. The development of the Renewable Electricity Directive in the early 2000s was not linked to any other policy process. Hence, there was not much room for bargained interaction. Even though it was part of the later climate policy package, work on the 2009 renewables directive does not appear to have been involved, in any significant horse-trading with other issue-areas.

Finally, we might expect to find influence from the energy policy for buildings through the mechanism of institutional interaction, since both policies have an intrinsic technology-development character. However, this similarity is better explained by similarities in the mind-set of Commission officials from DG Energy who initiated both policy areas. Nor do we see any institutional spill over between the ETS and renewable energy policy, even though the ETS produced more tangible consequences from around 2005 than did the policy on renewable energy. Instead, the renewable energy policy outcome in 2009 had less of a market flair than the 2001 outcome.

The renewables case contains a very interesting example of entrepreneurial interaction. Commission officials and electricity industry representatives portrayed the ETS design as a guide for the development of EU policy on renewable energy, and tried to persuade EU member states to opt for a Single European Market approach. These market entrepreneurs were persistent in their persuasion efforts, producing increasingly elaborate proposals for a pan-European green certificates scheme. Despite devoting considerable energy to this, they did not succeed. In a way, they were fighting an uphill battle: they lacked good examples of successful market measures in renewables, and were opposed by renewables actors.

These market entrepreneurs may have failed badly in inducing market thinking in the renewables policy, but they made a significant contribution to greater centralization: they succeeded in creating a key for national target distribution and flexible mechanisms that at least opened for cross-national collaboration. This kind of entrepreneurial interaction is not really captured by the three other interaction mechanisms. Bargained interaction involves tactical linking of issues, whereas this interaction had more of an institutional tinge: the ETS design was used to boost the legitimacy and appropriateness of certain proposals for renewable energy policy. This is what we call *persuasion entrepreneurship*. However, despite such efforts at persuasion entrepreneurship, the renewables outcome was not much affected by interaction. All in all, the character of policy here was shaped largely by issue-internal dynamics.

Turning to *CCS policy* we find a radically different picture. Here the policy drive emerged primarily as part of the broader EU climate policy drive in 2005. Looking through functional interaction lenses, one could argue that the volatile and generally rather low carbon price created the 'need' for an additional funding mechanism. However, it was not obvious that the need should be met. But the prospects of ETS auctioning revenues and the very set-up of the climate package, with different climate measures negotiated at the same time, formed a crucial and 'handy' basis for the interaction that played out in the course of 2008. Yet, these favourable background conditions would probably not have affected the policy outcome if the CCS entrepreneurs had failed to seize the opportunity and had not initiated bargained interaction.

So such bargained interaction proved essential in this case. The idea of linking ETS and CCS policy came from civil society actors and was facilitated and developed by the two Parliamentary Rapporteurs for CCS and the ETS, respectively. As to its essential form, the link developed more through collaboration and dialogue than horse-trading between the issue-areas in question. Member states played only a marginal role in the linking efforts, although the final outcome of the process (as to the exact number of ETS allowances to be set aside for CCS) also hinged on Council bargains. This bargained interaction is particularly crucial when it comes to explaining the competence distribution here: because the ETS was developing a high degree of centralization this occurred in relation to CCS as well.

Turning to institutional interaction, we do not see any instances of this. Despite the link to the ETS, the CCS outcome was flavoured by another institutional logic: the ETS is a market instrument, while the CCS policy is a technology-development instrument. There is nothing to indicate that the technology development flavour of CCS came about due to influence from other EU policy areas.

The main conclusion is still that important elements of the EU CCS policy emerged as a consequence of interaction. The strong centralization of the CCS policy outcome was to a significant degree shaped by bargained interaction, and to a certain extent underpinned by functional interaction. We find no significant institutional influence from other policy areas. The driving forces for the actual steering method must be located internally, within this issue-area.

Finally we turn to *the EU energy policy for buildings*. As noted in connection with the case of renewable energy, functional interaction did not shape this case much in the 1980s and 1990s. After the turn of the century, one might expect that the renewable energy target – which depends partly upon energy efficiency, as the share of renewable energy increases if the energy efficiency improves – contributed to the slight increase in support for an EU energy policy for buildings. Since the member states continued to reject strong centralization in this issue-area, this increase was not of any great importance. Moreover, the first directive on renewable energy served to strengthen the renewable energy industry, making it more able to influence the EU energy policy for buildings. This probably helped to strengthen the renewable energy content of the buildings policy, but without affecting the steering method or competence distribution to any extent.

With respect to bargained interaction, the renewable energy industry managed to create a link between the renewable energy directive decisions in 2009 and the EPBD decision in 2010: the new renewables directive included formulations on the use of renewable energy in buildings. Even though this served to strengthen the renewable energy focus of the EPBD somewhat, it did not change the competence distribution or the steering method. Finally, we do not see much institutional interaction in this process; as discussed earlier, renewable energy and the energy policy for buildings both had a technology-development steering method, but this did not stem from interaction.

We conclude that this policy area was very little shaped by interaction, apart from some slight functional interaction. The policy outcome developed first and foremost according to its own internal logic.

Theory Implications: Entrepreneurial Interaction has Increased Centralization of Control

Although the four policies are central elements of EU climate policy, and have on several occasions been discussed in conjunction, policy interaction has not really been important for the emergence and character of the policies. Initially, we expected the EU policies on CCS and on buildings to be fairly heavily influenced by interaction, and that we would see less interaction in the two other cases. This proved correct with respect to CCS, but the energy policy for buildings turned out to be less influenced by interaction than we had expected. Renewable energy was slightly influenced by interaction. Despite the rather modest interaction imprint, our cases offer instances of interesting variation, and enable us to formulate some conclusions with implications for theory.

The ETS stands out as the main policy sender. It was intended to serve as the flagship of EU climate policy, so other climate policies that affect the same sectors (such as renewables and CCS) can easily be seen as interfering with the carbon pricing mechanism. The fact that these other policies have been strengthened may indicate that the market thinking of the ETS has not really been institutionalized. Not even the active interaction entrepreneurship conducted in the renewable energy case acted to diffuse the market thinking.

Without doubt, interaction is complex, hard to capture analytically. Our case studies show that, to understand how various policies affect each other, there is a need for four interaction mechanisms – not three, as we suggested initially. In Table 8.5 we show how the four interaction mechanisms can be categorized in relation to two underlying dichotomies, the mechanism type dimension and the power source dimension.

Table 8.5   Four EU climate policy interaction mechanisms

Mechanism type	Power source	
	Structural	*Institutional*
Social	1) Functional interaction	2) Institutional interaction
Entrepreneurship	3) Bargained interaction	4) Persuasive interaction

Functional interaction is a social interaction, in the sense that it is a result of how a certain policy is implemented and functions, and not how it has been initiated by any specific entrepreneurs. However, it is structural because it responds to shifts in the distribution of authority and information (see Boasson 2011 for a more thorough discussion on the basis of structural power). In our cases, this mode of interaction played a marginal role, but we do not think that this renders this mechanism unimportant. Rather, we suggest that functional interaction has had limited importance thus far simply because considerable time has been needed for EU climate policies to move beyond mere symbolism. Obviously, policies without much substance cannot produce consequences of functional interaction. With the recent strengthening, more functional interaction can be expected in the future.

Also institutional interaction is social, but it pertains to how certain policies can create shifts in norms and mind-sets in other areas. This mechanism proved to be of little importance in our cases. This may mean that because the different policies are embedded in different organizational fields, we will not find much cross-fertilization in the norms and institutional logics that underpin the policies. However, in the future the EU may well develop stronger coordination between issue-specific climate policy development processes, which could serve to undermine the mechanisms specific to the organizational field.

Bargained interaction refers to interaction initiated by entrepreneurs who seek to increase their structural power basis in one policy area by linking it to another. Through bargained interaction with the ETS, the CCS entrepreneurs were able to ensure that the CCS outcome became far more centrally governed than would otherwise have been the case. This is the most important interaction in our case sample. Brussels-based lobbyists and MEPs, not member states, were the prime initiators of bargained interaction.

In the ETS case we find some instances of nationally initiated policy linkages, but this is the exception to the dominance of Brussels-situated actors, and runs counter to the LI view that national governments will be the most important initiators of interaction (Moravcsik 1993, 1998). Our findings are hence more in line with the thinking of MLG scholars (Hooghe 2001, Hooghe and Marks 2001, Backe and Flinders 2004), although they pay little attention to interaction as such. We hope that our conclusions can encourage MLG scholars to explore the extent to which EU policy entrepreneurs initiate interaction to strengthen their impact.

Furthermore, we detected a fourth kind of interaction in the renewables case: what we have termed persuasion interaction. This interaction is initiated by policy entrepreneurs deliberately seeking to create normative and cognitive shifts in one issue-area by linking it to another area. As with bargained interaction, this mechanism is initiated by entrepreneurs – but here entrepreneurs actively try to transfer a policy model from one issue-area to another, arguing that the other policy has superior qualities. The Commission attempted persuasive interaction in the renewables case in order to ensure outcome similarity between the ETS and renewables. They did not succeed in spurring the creation of a Single European Market outcome, but they did contribute to greater centralization within the issue-area.

Sociological NI studies on the diffusion of organization recipes have found interaction effects similar to the persuasion interaction mechanism. Policy linking can be a way to show that a certain idea is desirable, proper or appropriate (Scott 2008: 59). This can even be like following the latest fashion: the interaction initiator aims for conformity with dominant policy recipes (Sahlin and Wedlin 2008: 223). The challenge is to link a policy proposal to a policy that actually has widespread legitimacy, and here the Commission failed. Since the Single European Market had not been set in operation yet in the case of the ETS in 2008/2009, the ETS was not able to fully function as a legitimizing template for a Single European Market-type of policy on renewable energy. Moreover, we find it reasonable to expect that the character of the EU climate policy that comes to be perceived as the most successful will diffuse to other issue-areas, whether through institutional interaction or persuasion interaction. If the revised ETS becomes more of a success, managing to create a significantly high and stable price of CO_2, will be particularly important.

On this backdrop, we conclude that the two instances of interaction with effects on the outcomes – bargained interaction in the CCS case and persuasion interaction in the renewables case – have both led to increased centralization of EU climate policy. Both instances have been of the entrepreneurial type. Hence, our main interaction conclusion is that thus far, *policy interaction has contributed slightly to increase the centralization of control in EU climate policy.* This finding can be seen as being in line with new-functionalist arguments, but the overall effect is far weaker than predicted by this school (see Haas [1958] 2004, Niemann 2006). We do not detect any systematic relationship between policy interaction and steering method. However, we have noted several factors indicating that social policy interaction mechanisms may become more important in the future, perhaps also affecting the steering method of climate polices. On the other hand, this is not to say that we believe that EU climate policy will necessarily become more coherent in the future.

We must warn against underestimating the cognitive and collaborative challenges involved in creating of a coherent EU climate policy. Oberthür and Gehring (2011: 48) argue that in order for more deliberate policy interaction to play out, the policymakers must 'take into account the broader policy implications' of the particular governance project. This is no easy task where EU climate policy is concerned. Indeed, it is not even clear what a coherent or integrated EU climate policy should or could look like. For example, Helm (2009) maintains that the 20-20-20 targets lack internal coordination, and that various EU climate policies are partly in conflict with each other.

Our case studies show that the Commission aimed for greater harmonization in the sense that climate policies were to have more of a market character. But similarity in steering method is no guarantee of harmonization. Had the Commission succeeded in creating a Single European Market both for renewables certificates and for carbon allowances, two different markets would have emerged. The pricing mechanisms of the two markets would probably have affected each other, but it is very hard to visualize how and to what degree this would have happened – not to mention how to regulate them in order to ensure a coherent and harmonized climate effect.

In addition to the intellectual challenges involved, no single actor is sufficiently dominant to control the totality of EU climate policy development. Most issues develop in line with issue-specific dynamics. The overall limited interaction can be seen as a result of organizational barriers between the various Commission DGs and between different policy areas, as indicated by public administration studies of the EU executive (Egeberg 2006a, Trondal 2010). True, a separate Climate DG has recently been created within the Commission and this might make it easier for the Commission to initiate linkages between policy areas. However, this DG has limited power: important EU climate polices, such as renewables and energy efficiency, are still managed by DG Energy.

Finally, our assessment indicates that the interaction efforts that actually play out are not aimed at increasing the coherence of the climate policies. The important interaction entrepreneurs are actors without responsibility for the coherence of EU climate policy, such as Brussels lobbyists and MEPs. For instance, the bargained interaction in the CCS case occurred because it was convenient for the *carpe diem* entrepreneurs, not as a means to increase the coherence of EU climate policy.

External Environment: Enabling Entrepreneurs to Create Windows of Opportunity

Case Comparison: Primarily International Institutional Entrepreneurship

Here we assess to what extent and how the emergence and character of climate policy outcomes have been shaped by:

- *Coercive Pressure:* international market conditions or global regime obligations (LI)
- *Institutional Diffusion:* norms and policy recipes adopted and promoted by other countries and international organizations (NI)
- *International Entrepreneurship:* entrepreneurs making strategic use of international networks and windows of opportunity (MLG).

Table 8.6   How external mechanisms shaped the policy outcomes – rough scores

External environment mechanism	ETS	Renewables	CCS	Buildings	Overall importance of mechanism
Coercive pressure	Little	Little	Little	Little	Little
Institutional diffusion	Some	Little	Some	Little	Some
International institutional entrepreneurship	Significant	Significant	Significant	Some	Significant

Table 8.6 summarizes and compares the importance of our mechanisms for the character of the policy outcomes in the three cases. Taking stock of effects from outside the EU (the external environment), we begin with the *ETS* case. Important decisions on EU emissions trading policy have correlated with main developments and milestones in the global climate regime. This pattern has occurred at least twice: around the turn of the millennium and in 2007–2009.

Did this correlation result from coercive pressure? The concept of emissions trading was taken up within the global climate regime *before* the EU started to develop a system of its own, although the Kyoto Protocol did not formally *require* the development of an EU ETS. Furthermore, related to the lack of the development of a global carbon price, we do not see any significant market changes. Hence, coercive pressure does not seem to have contributed much to the emergence of the EU ETS. Still, increasing attention given to the possibility of global competitive and market effects has underpinned recent calls for exemptions to the Single European Market outcome.

What about institutional diffusion? First, the adoption of the 1997 Kyoto Protocol and not least its flexibility mechanisms served as an important stimulus for the EU's subsequent volte-face and the development of an ETS. Hence, developments within the global regime contributed to legitimize emissions trading and made this policy recipe appear more attractive to EU policymakers. While this may help to explain the actual adoption of the ETS, it cannot explain why the EU later streamlined and centralized the system; at that time, few other important global actors had adopted emissions trading.

Can international entrepreneurship help to explain the development of EU policy? Yes, we certainly note that entrepreneurs repeatedly exploited critical junctures in the global regime to facilitate EU climate action. Not least, the Commission used the US withdrawal from the Kyoto Protocol in March 2001 to create such a window of opportunity. It was argued that in order to bring the international process forward, the EU would have to step up the work on developing internal policies, the ETS in particular. Furthermore, the new EU climate and energy targets and policy package put forward in 2007 and 2008 were cast as means and instruments for achieving an ambitious and comprehensive agreement at the Copenhagen climate summit in December 2009.

The international entrepreneurship of the Commission had a strong institutional character, and this contributes to explain why many actors came to regard a centralized, streamlined and smoothly functioning ETS as a sheer necessity. Commission entrepreneurs also succeeded in convincing EU policymakers that such an ETS could help to bolster EU ambitions for global leadership, while strengthening the EU's ability to comply with anticipated, more ambitious post-Kyoto global rules. Hence, international entrepreneurship emerges as instrumental in framing and polishing the ETS as the flagship for a global crusade. This mechanism contributed to greater centralization as well as to the market streamlining of the ETS measure.

Turning to *renewable energy*, also here we note that important EU renewable energy policy decisions correlated with main decisions in the global climate regime. This pattern appears three times: in the early 1990s, at the entrance to the 2000s and in 2007–2009. Did this result from coercive pressure? That seems highly unlikely, since neither the Climate Convention nor the Kyoto Protocol included specific, binding renewable energy commitments. Moreover, in the early 1990s and in 2007–2009, the EU adopted more stringent policies based on the anticipation of global decisions, not as a result of actual global decisions. True, the initiation of the EU renewables policy came in response to global market changes – the oil crisis in the 1970s. Yet, it was not obvious that the EU as such should respond, and not only its member states. Also this response hinged on how EU actors chose to interpret external developments.

Did institutional diffusion play a role here? The answer is clearly no: the global climate regime did not promote a technology-development approach to renewable energy, and there was nothing in the EU-external environment to indicate that it was the EU and not the member states that should take the lead in this policy development.

Turning to international entrepreneurship, we see similarities to the ETS case: Commission officials used the situation in the global climate negotiations to create EU-internal windows of opportunity. This was especially clear in 2008, as certain member-state representatives and Commission officials were able to create an understanding of urgency, framing a swift development of ambitious EU policy as instrumental for the development of a new global climate deal. Also the renewable energy industry made frequent use of this rhetoric.

As we saw in the ETS case, this entrepreneurship had a strong institutional flair, with entrepreneurs using the global backdrop to alter how EU policymakers perceived the decision situation. Hence, we conclude that international institutional entrepreneurship played an important role with respect to the development of the EU policy on renewable energy. While the external environment provided little impetus for a technology-development approach, it contributed to strengthen the degree of centralization.

Due to the short history of *CCS policy*, we can only assess how it correlated with the global climate regime from 2005 onwards. The EU started to develop a CCS policy only after CCS had emerged as an issue in various international climate forums and the international energy security situation had become increasingly strained. Yet, none of these developments resulted in binding CCS commitments for the EU. Moreover, the changing conditions in the international energy market could be countered by various measures, of which CCS was merely one. Hence, the policy was not much affected by coercive pressure. However, institutional diffusion does seem to have been in operation. The growing involvement of credible, international research-based organizations made CCS stand out as a more legitimate and promising technology for climate-change mitigation.

These international developments also paved the way for entrepreneurship. As seen in the other cases, the feeling of urgency surrounding the development of the climate package helped to legitimate the development of an ambitious EU CCS policy. But while Commission officials devoted little entrepreneurial energy to CCS specifically, other actors moved to the fore to exploit this opportunity to promote CCS. Most importantly, this can be seen as creating the special background conditions for the entrepreneurship of the Parliament Rapporteurs and industry representatives. In particular, the entrepreneurs exploited this situation to induce fairly strong centralization of control here.

Finally, how did the external environment affect *the EU energy policy for buildings*? The timing of decisions is very similar to the case of the policy on renewable energy. In addition, neither the climate regime nor soaring energy prices provided coercive pressure on the EU to develop an energy policy for buildings. Concerning institutional diffusion, we cannot see that international organizations or other countries led the way for an EU local loading outcome. True, certification of buildings had been developed in the USA at about the same time as in the EU, but this did not serve as an important impetus for the development of the EU policy in this area.

However, the external environment did affect policymaking somewhat, through the same international institutional entrepreneurship seen in the other cases – in particular, by ensuring that the issue was repeatedly put on the EU agenda. But in the last phase, from 2005 to 2010, neither the Commission nor other actors engaged themselves with the same entrepreneurial intensity as in the other cases. Hence, we conclude that the case of energy policy for buildings was less affected by the external environment than were the other policy areas.

Theory Implications: Strengthening the Commission and Leading to Climate Policy Emergence

In line with expectations, we find that the external environment was crucial for the ETS and CCS. Moreover, buildings and renewables were more affected than we had initially expected. The external environment contributed to explain the emergence of all the policies, but while such factors were important drivers for increased centralization in the cases of the ETS, renewables and CCS, they had only moderate effect on energy policy for buildings. Moreover, the external environment contributed to shape the steering method of the ETS, but not the other policy areas.

The external environment exerted less direct influence than predicted by Liberal Intergovernmentalism. In all cases we find little coercive pressure, so it seems reasonable to conclude that this mechanism played only a minor role. However, the picture changes somewhat when we recall that, while the Kyoto Protocol did not require the EU to develop any specific climate measures, it *did* strongly induce the EU to introduce some actions aimed at reducing emissions (Oberthür and Tänzler 2007). Hence, it can be argued that the EU was 'forced' to adopt some climate measures, although it was up to the EU policymakers to choose and design the specific measures.

EU policy has tended to emerge and change in the run-up to key international climate decisions, and this indicates that anticipatory effects may have been more important than actual pressures. Can this be explained by institutional diffusion, with peer pressure and international diffusion of policy recipes? We have seen that international organizations promoted market-based ETS and technology-development CCS policy, and these served as sources of inspiration for EU actors in the two cases. That was not the situation for renewable energy and energy policy for buildings.

Moreover, none of the four policies were firmly pursued by many countries outside the EU. After the US withdrawal from the Kyoto Protocol, the EU had no suitable international peer. CCS is the only issue-area in which the USA developed tangible policies before the EU. In general, we do not see much institutional diffusion; only to a very small extent could the EU copy what others had done – in general, it had to draw on internal resources in order to create climate policies. Hence, institutional diffusion does not stand out as the mechanism best suited to capturing the influence of the external environment.

Did the international influence then take place primarily through international entrepreneurship? There was considerable entrepreneurial activity related to the external environment, but not of the type that we expected. In line with MLG thinking, we initially suggested that international entrepreneurs would make strategic use of international networks and windows of opportunity. However, external developments did not produce 'objective shocks and crises', interpretation was key. Neither do we find much strategic networking outside the EU. The international entrepreneurship had an institutional character, consisting of two components: entrepreneurs used developments in the climate negotiations to *create* windows of opportunity, and to increase the perceived legitimacy of their favoured climate policy recipes.

Within the field of political science, the understanding of 'windows of opportunity' builds heavily on John Kingdon's studies of US federal policymaking (Kingdon [1984] 2011). According to Kingdon, whether and how well such windows are exploited will depend on the activities of entrepreneurs. This is not only a question of 'objective features' but exists 'in the perceptions of the participants as well' (Kingdon [1984] 2011: 171). However, he did not specify just how entrepreneurs could go about changing actors' perceptions of certain decision situations.

In Chapter 3 we saw how the Commission, sometimes with help from key member states, framed decision situations in the international climate regime as policy windows. For instance, from 2005 and onwards, Commission officials contributed to a cognitive framing along the lines that 'the EU can contribute to solve the global negotiation hurdle by developing ambitious EU-internal policies.' Further, when the Council in 2005 called for stronger EU climate policies, the Commission responded by creating an extraordinary procedure, starting with target setting, followed by a whole batch of policy proposals decided upon in a remarkably swift and efficient decision-making procedure. Germany and France, which held the EU Presidency at crucial moments, also supported and contributed to creating executive procedures that opened and extended the policy window.

The policy window opened up for considerable specific entrepreneurial activity directed at issues of climate policy package, such as the ETS, renewables and CCS. This probably contributed to the increased centralization of these issues. The buildings case was to a lesser extent affected by the need for radical decisions before the Copenhagen summit. It appears to have helped to ensure that member states agreed on an outcome, but did not foster stronger centralization. Here we should recall that international institutional entrepreneurship had played out earlier as well – for instance, it had contributed to the emergence of policies for renewables and buildings in the 1990s.

In addition, international institutional entrepreneurs used international developments to reinforce the legitimacy of proposing and developing EU policies on emissions trading and CCS. But their legitimating technique did not involve arguing that the EU should follow other actors, as described by Finnemore and Sikkink (1998). It was more in line with 'fashion following', as this is described by sociological new institutionalists (see Sahlin and Wedlin 2008). Adoption of the ETS and CCS were in line with dominant international trends, but the EU policies were also pitched to be somewhat better than what the others were doing. A main reason for the adoption of these policies was that the EU aimed to become a global climate leader – or the 'fashion queen' of international climate policy (see Boasson 2011 for a discussion on fashion queen entrepreneurship). Since renewable energy and energy policy for buildings attracted less international attention, this entrepreneurial technique was not as important to the final character of these policies.

We conclude that the *external environment primarily affected EU climate policymaking through international institutional entrepreneurship.* Due to its executive role, the Commission was particularly well-placed for performing such entrepreneurship. Moreover, the EU member states had a common Kyoto Protocol commitment. The external environment perspective complement and nuance the literature on the Commission's ability to set the substantive agenda for the EU. Hitherto, the discussion has focused largely on how the Commission can increase its power under conditions of imperfect information, while EU external factors have received scant attention (see Pollack 1997: 126, Hix 2005: 68).

But the Commission was not the only international institutional entrepreneur. In 2005 and the following years the environmental movement, the renewable energy industry and certain national political leaders, such as Merkel, Blair and Sarkozy, contributed as well. These national leaders seem to have risen to the occasion and executed what we have termed *carpe diem* entrepreneurship: sensing that climate change had become 'high politics', they grasped the chance to show leadership. In contrast, the Commission was more of a slow and steady 'tortoise'. At every juncture of the international climate talks, they kept on exerting international institutional entrepreneurship. It was first and foremost these repeated efforts of the Commission that ensured that the global climate talks came to set the pace of the EU climate policy development.

In general, the international institutional entrepreneurship served to expand the room for network entrepreneurship and bargained interaction entrepreneurship in the issue-specific climate policy areas. As suggested by Kingdon, advocates of various proposals sensed their opportunity and rushed to take advantage of it (Kingdon [1984] 2011: 175). In some instances, tortoise entrepreneurs had been waiting for an entrepreneurial opportunity and proved clever at exploiting it when it arose – this applies to 'Commission officials' in the ETS case, the renewable energy industry in the renewables case, and the insulation industry in the buildings case. As noted, the CCS case was primarily characterized by *carpe diem* entrepreneurship.

Our second conclusion is that *the external environment primarily contributed to policy emergence and policy centralization.* The key role of the Commission as international entrepreneur and the heightened Brussels entrepreneurship that followed from the policy windows both contributed to centralization. This is clear in the cases of the ETS, renewables and CCS. The centralization effect was marginal in the case of energy policy for buildings, particularly because centralization was undermined by the decentralized field structure in this policy area. To some extent the external environment affected the steering method of the ETS, but this effect was weaker than it was with respect to centralization.

We would probably have seen more variation in external environment mechanisms outside the EU if more global actors had implemented ambitious climate policies and if the climate negotiations had resulted in more stringent obligations. It seems rather unfounded to expect radical changes in the climate regime anytime soon; hence, international institutional entrepreneurship is likely to remain the dominant mechanism through which the EU-external environment will affect EU climate policy also in the future.

Conclusions

This chapter has discussed how we best can explain EU climate policy development, and presented conclusions based on our historical examination and comparison of the four cases of EU climate policy chosen for this study. The external environment has emerged as by far the most important contributor to policy emergence, providing key explanations for the emergence of all four policies. Industry influenced the emergence of the ETS, but not the others. Interaction contributed only to the emergence of CCS policy. When it comes to explaining policy characteristics, we have noted several different causal patterns. All policy outcomes were influenced by industry. Policy interaction was crucial for the character of the CCS policy and had some importance for the character of the policy on renewables, but was otherwise not significant. Finally, developments outside the EU contributed to increase the centralization of the ETS, renewables and CCS and also influenced the steering method of the ETS.

What were the mechanisms at work? All cases are best explained by a combination of institutional feedback from the organizational fields in which the policy issues are embedded and various kinds of entrepreneurship: network entrepreneurship, bargained entrepreneurship and international institutional entrepreneurship. Industry influenced the policy outcomes through a combination of institutional feedback and network entrepreneurship. The Commission, politicians from some dominant member-states, but also some industry actors exercised international institutional entrepreneurship. The Parliament and non-governmental actors increased their influence through bargained interaction.

Moreover, we are now able to clarify the puzzling differences among the policy outcomes. First, differences in the institutional feedback mechanisms provide the key explanation for why we find a shift towards centralization in the ETS, renewables, and CCS cases, but not in the case of energy policy for buildings. International institutional entrepreneurship comes in addition, reinforcing the field-specific effects in the three first-mentioned cases. (And here we note that the high degree of centralization in the CCS case was the result of bargained interaction with the ETS.) Second, the special institutional feedback mechanisms and international institutional entrepreneurship in the ETS case explain why this policy is characterized by a steering method unlike that of the other issue-areas.

Our focus has been on explaining EU climate policy as of 2010. However, our findings can also be used to develop more general propositions that can be used to analyse and explain other EU climate policies than those examined in this book, and for reasoning about future climate policy developments. Our main conclusions offer a basis for developing of more general theory propositions.

A key finding is that industry has been a much less of an obstructive braking block than anticipated, and has generally played a quite nuanced and multi-faceted role. We can derive three propositions as to how industry shapes climate policy development. First, we posit that the distribution of authority within the organizational field will tend to shape the competence distribution of outcomes. Our second industry thesis is that the institutional logic that dominates the organizational field will tend to infuse the steering method of the policy outcome. Third, we suggest that Europeanized organizational fields will produce industry entrepreneurship in centralized fields, rather than in decentralized ones. These suggestions indicate that in order to understand the industry impact it is important to take the structure and institutional logic of the larger organizational fields into account.

This means that the organizational field that specifically underpins a certain policy is of high importance. For instance, we have seen that the character of the EU's renewable energy policy is rooted in the organizational field of renewable energy. Actors in the organizational field of European electricity production have also tried to exert influence, but have had scant impact so far. On the other hand, developments within the two fields may lead to a shift from the renewables field to the field of electricity production, probably with major ramifications for the steering method. There is no reason to expect similar changes for the ETS and the energy policy for buildings. CCS was rooted in the field of petroleum production, but later the electricity field embraced it. Thus far, the issue has not become deeply embedded in any of the fields. The dynamic of future policy development here will depend on to what extent and how CCS becomes embedded in the two fields.

Concerning interaction, we see fewer ramifications for most issue-areas, with CCS as the exceptional case. While the making of 'policy packages' certainly helped create a new policy drive after 2005, it brought fewer interaction effects than the 'integrated climate package' rhetoric would lead us to believe. Most policies have developed mainly due to issue-internal factors. Yet, our assessment allows us to offer one proposition here: that policy interaction tends to contribute to increase the centralization of EU climate policy. This finding is in line with new-functionalist arguments, even though the effect is of less importance than predicted by this school (see Haas [1958] 2004, Niemann 2006).

We do not detect any systematic relationship between policy interaction and steering method, but this does not imply that such systematic relationships may come into play in the future. Moreover, we posit that interaction will play a larger role in the future development of EU climate policy. The extent of this will depend on the success of existing climate policies as well as developments in the global climate regime. After all, how long can the EU continue to maintain strong climate policies if the rest of the world does not follow suit?

More than often recognized, the external environment has helped set in motion and fuelled the development of EU climate policy, but instead of exerting coercive pressure, these international factors have acted through entrepreneurs. Here we derive two theory propositions from our case studies. First, we propose that the external environment will affect EU climate policymaking primarily through international institutional entrepreneurship. This is largely a reflection of the weak international climate regime. Since it does not seem that a more binding regime is likely to appear anytime soon, we have no reason to expect that other external mechanisms will become more important in the near future.

However, we do see tendencies for big actors, such as China, to adopt more ambitious climate policies, even without binding international commitments. That means that the EU may come to experience tougher peer competition, which in turn may serve to increase the importance of the mechanism of institutional diffusion. Second, we have suggested that the external environment will contribute to increase the centralization of the EU climate policies. However this tendency is dependent upon the continued dominance of the mechanism of international institutional entrepreneurship. Moreover, any major changes in the external environment will probably lead to shifts in its impact on competence distribution and steering method.

Finally: we hope that this book, and not least the mechanisms identified and tested here, have managed to create a window of opportunity for more scholars to jump in and execute analytical *carpe diem* entrepreneurship on the understanding of EU climate policy. We are confident that giving prime attention to industry, policy interaction and the external environment will provide important causal clues in such ventures, but this should in no way block the discussion of a wide range of other factors. EU climate policy has been and will remain a complex, multi-level 'grand experiment'.

References

Acrill, R. and A. Kay, 2005. Multiple streams in EU policy-making: the case of the 2005 sugar reform. *Journal of European Public Policy*, 18(1), 72–89.

Andresen, S. and Boasson, E.L. 2012. International climate cooperation: clear recommendations, weak commitments, in *International Environmental Agreements – An Introduction*, edited by S. Andresen, E.L. Boasson and G. Hønneland. London: Routledge, 49–67.

Andresen, S., Boasson, E.L., and Hønneland, G. (eds.) 2012. *International Environmental Agreements – An Introduction*. London: Routledge.

Bache, I. and Flinders, F. 2004. Themes and issues in multi-level governance, in *Multi-level Governance*, edited by I. Bache and F. Flinders. Oxford: Oxford University Press, 1–11.

Barnes, P. 2011. The role of the Commission of the European Union: creating external coherence from internal diversity, in *The European Union as a Leader in International Climate Change Politics*, edited by R.K.W. Wurzel and J. Connelly. Oxon: Routledge, 41–58.

Bellona 2008. Joint open letter from Alstom, Bellona, E3G, Climate Change Capital, Fortum, Shell, Sintef and Vattenfall to EU energy and environment ministers. Brussels, 30 June 2008. [Online]. Available at: http://bellona.org/filearchive/fil_CCS_leadership_letter_to_ministers_30June08.pdf [accessed: 8 February 2012].

Bellona 2010. Coalition urges EU ministers to stimulate CCS. [Online: Bellona]. Available at: www.bellona.org/articles/articles_2008/ccs_coalition [accessed: 13 August 2011].

Bennett, A. 2008. 'Process Tracing: a Bayesian Perspective', in *The Oxford Handbook of Political Methodology*, edited by J.M. Box-Steffensmeier, H.E. Brady and D. Collier. Oxford: Oxford University Press, 702–21.

Benson, D. and Jordan, A. 2010. The expansion of EU climate policy and its future under the Lisbon Treaty. *St Antony's International Review*, 5(2), 121–40.

Boasson, E.L. 2005. *Klimaskapte beslutningsendringer? En analyse av klimahensyn i petroleumspolitiske beslutningsprosesser*. FNI Report 13/2005. Lysaker, Norway: Fridtjof Nansen Institute.

Boasson, E.L. 2009. On the management success of regulative failure: standardised CSR instruments and the oil industry's climate performance. *Corporate Governance. The International Journal of Business in Society*, 9(3), 313–25.

Boasson, E.L. 2011. *Multi-sphere Climate Policy: Conceptualizing National Policy-making in Europe*. PhD thesis. Oslo: Department of Political Science, University of Oslo.

Boasson, E.L., Bohn, M. and Wettestad, J. 2009. CSR in the European oil sector: a mapping of company perceptions, in *Corporate Social Responsibility in Europe. Rhetoric and Realities*, edited by R. Bart and F. Wolf. Cheltenham: Edward Elgar, 65–79.

Bourdieu, P. 1992. The practice of reflexive sociology', in *An Invitation to Reflexive Sociology*, edited by P. Bourdieu and L.J.D. Wacquant. Cambridge: Polity Press, 216–60.

Box-Steffensmeier, J.M, H.E. Brady and David Collier. 2008. 'Political Science Methodology', in *The Oxford Handbook of Political Methodology*, edited by J.M. Box-Steffensmeier, H.E. Brady and David Collier. Oxford: Oxford University Press, 3–31.

BP 2009. Capturing carbon dioxide. [Online]. Available at: www.bp.com/ sectiongenericarticle.do?categoryId=9023211&contentId=7043026 [accessed: 5 February 2010].

Buhr, K. and Hansson, A. 2011. Capturing the stories of corporations: a comparison of media debates on carbon capture and storage in Norway and Sweden. *Global Environmental Change*, 21, 336–45.

Build up 2010. Energy performance of buildings: the Commission asks Italy and Spain to ensure full compliance with European legislation. [Online]. Available at: http://buildup.eu/news/12614 [accessed: 8 February 2012].

Burns, C. and Carter, N. 2011. The European Parliament and climate change: from symbolism to heroism and back again, in *The European Union as a Leader in International Climate Change Politics*, edited by R.K.W. Wurzel and J. Connelly. Oxford: Routledge, 58–74.

CAN, WWF, Greenpeace, Friends of the Earth. 2007. ECCP EU ETS review process: written comments CAN-Europe, Friends of the Earth Europe, Greenpeace and WWF. Brussels, June.

Carbon Trust. 2008. *Cutting Carbon in Europe – The 2020 Plans and the Future of the EU ETS*. London: Carbon Trust Publication CTC734, June.

Cass, L. 2005. Norm entrapment and preference change: the evolution of the European Union position on international emissions trading. *Global Environmental Politics*, 5(2), 38–60.

Cass, L. 2007. The indispensable awkward partner: the United Kingdom in European climate policy, in *Europe and Global Climate Change. Politics, Foreign Policy and Regional Cooperation*, edited by P. Harris. Cheltenham: Edward Elgar, 63–87.

Center for Public Integrity 2009. *The EU's Billion-Euro Bet: How Europe Ended Up Paying Industry for Carbon Capture and Storage* [Online]. Available at: www.publicintegrity.org/investigations/global_climate_change_lobby/ articles/entry/1860/ [accessed: 8 February 2012].

CEN (European Committee for Standardization) 2009. Energy performance of buildings. Available at: www.cen.eu/cenorm/sectors/sectors/construction/ sustainableconstruction/epbd.asp (accessed: 21 January 2011).

Christiansen, A.C. and Wettestad, J. 2003. The EU as a frontrunner on greenhouse gas emissions trading: how did it happen and will the EU succeed? *Climate Policy*, 3(1), 3–18.

Claes, D.H. and Frisvold, P. 2009. CCS and the European Union: magic bullet or pure magic?, in *Caching the Carbon – The Politics and Policy of Carbon Capture and Storage*, edited by J. Meadowcroft and O. Langhelle. Cheltenham: Edward Elgar, 211–35.

Clémencon, R. 2008. The Bali road map: a first step on the difficult journey to a post-Kyoto protocol agreement. *The Journal of Environment and Development*, 17, 70.

Codognet, M.K., Glachant, J.M., Hiroux, C., Mollard, M., Lévêque, F. and Plagnet, M.A. 2003. *Mergers and Acquisitions in the European Electricity Sector*. Report. Paris: Centre d'economique industrielle.

Coen, D. 1998. The European business interest and the nation state. *Journal of Public Policy*, 18(1), 75–100.

Coen, D., Grant, W. and Wilson, G. 2010a. Overview, in *The Oxford Handbook of Business and Government*, edited by D.W. Coen, W. Grant and G. Wilson. Oxford: Oxford University Press, 1–5.

Coen, D., Grant, W. and Wilson, G. 2010b. Political science: perspectives on business and management, in *The Oxford Handbook of Business and Government*, edited by D. Coen, W. Grant and G. Wilson. Oxford: Oxford University Press, 9–34.

Coen, D. and Richardson, J. 2009a. Introduction, in *Lobbying the European Union: Institutions, Actors and Issues*, edited by D. Coen and J. Richardson. Oxford: Oxford University Press, 3–15.

Coen, D. and Richardson, J. 2009b. Institutionalizing and managing intermediation in the EU, in *Lobbying the European Union: Institutions, Actors and Issues*, edited by D. Coen and J. Richardson. Oxford: Oxford University Press, 337–50.

Cohen, M.D, March, J.G. and Olsen, J.P. 1972. A garbage can model of organizational choice. *Administrative Science Quarterly*, 17(1), 1–25.

Collier, D. 1993. The Comparative Method, in *Political Science: The State of the Discipline II*, edited by A.W. Finifter. Washington, DC: American Political Science Association, 105–19.

Collier, D. and J. Gerring. 2009. *Concepts and Method in Social Science: The Tradition of Giovanni Sartori*. London: Routledge.

Collier, U. 1997. Sustainability, subsidiarity and deregulation: new directions in EU environmental policy. *Environmental Politics*, 6(2), 1–23.

Coninck, H. de and Bäckstrand, K. 2011. An international relations perspective on global politics of carbon dioxide capture and storage. *Global Environmental Change*, 21(2), 368–78.

Costa, O. 2011. Spanish, EU and international climate change policies: download, catch up, and curb down, in *The European Union as a Leader in International Climate Change Politics*, edited by R.K.W. Wurzel and J. Connelly. Oxford: Routledge, 179–95.

Czarniawska, B. andSevón, G. 1996. Travels of ideas, in *Translating Organizational Change*, edited by B. Czarniawska and G. Sevón. Berlin: Walter de Gruyter & Co., 13–48.

Dahl, R.A. 1961. *Who Governs? Democracy and Power in an American City*. Hew Haven, CT: Yale University Press.

Darkin, B. 2006. Pledges, politics and performance: an assessment of UK climate policy. *Climate Policy*, 6, 257–74.

Del Rio, P. 2007. Experiences from allocating allowances in the EU ETS: the case of Spain, in *Allocation in the European Emissions Trading Scheme: Rights, Rents and Fairness*, edited by A.D. Ellerman, B.K. Buchner, and C. Carraro. Cambridge: Cambridge University Press, 182–212.

DETR. 2000. *Delivering Emissions Reductions, Climate Change – The UK Programme*. London: Department of the Environment, Transport and the Regions.

Diez, T. and Wiener, A. 2009. Introducing the mosaic of integration theory, in *European Integration Theory*, edited by A. Wiener and T. Diez. Oxford: Oxford University Press, 1–22.

DiMaggio, P.J. and Powell, W.W. 1991. The iron cage revisited, in *The New Institutionalism in Organizational Analysis*, edited by W.W. Powell and P.J. DiMaggio. Chicago, IL: University of Chicago Press, 63–82. [first published in *American Sociological Review* 1983].

Dimitrov, R. 2010. Inside Copenhagen: the state of climate governance. *Global Environmental Politics*, 10(2), 18–24.

Downs, A. 1972. Up and down with ecology – the 'issue-attention cycle'. *Public Interest*, 28(Summer), 38–50.

Drori, G.S., Meyer, J.W. and Hokyo Hwang 2006. *World Society and Organizational Change*. Oxford: Oxford University Press.

Duwe, M. 2001. The Climate Action Network: a glance behind the curtains of a transnational NGO network. *RECIEL*, 10(2), 177–89.

Ecofys 2004. *Analysis of the National Allocation Plans for the EU Emissions Trading Scheme*. August. London: Ecofys UK.

EDF Group, 2009. *Annual Report 2008*. Paris Cedex: EDF Group.

Egeberg, M. (ed.) 2006. *Multilevel Union Administration*. Basingstoke: Palgrave Macmilllan.

Egelund Olsen, B. 2006. The IPPC permit and the greenhouse gas permit, in *EU Climate Change Policy – The Challenge of New Regulatory Initiatives*, edited by M. Peeters and K. Deketelaere. Aldershot: Edward Elgar, 153–69.

Eising, R. 2002. Policy learning in embedded negotiations: explaining EU electricity liberalization. *International Organization*, 56(1), 85–120.

Eising, R. 2004. Multilevel governance and business interests in the European Union. *Governance: An International Journal of Policy, Administration, and Institutions*, 17(2), 211–45.

Eising, R. and Kohler-Koch, B. 1999. Introduction: network governance in European Union, in *The Transformation of Governance in the European Union*, edited by B. Kohler-Koch and R. Eising. New York: Routledge, 3–13.

Ellerman, A.D. and Buchner, B.K. 2007. The European Union Emissions Trading Scheme: origins, allocation, and early results. *Review of Environmental Economics*, 1(1), 66–87.

Ellerman, A.D., Convery, F.J. and Perthuis, C.D. (eds). 2010. *Pricing Carbon – The European Union Emissions Trading Scheme*. Cambridge: Cambridge University Press.

Ellerman, A.D., Joskow, P.L., and Harrison, D.L. 2003. *Emissions Trading: Experience, Lessons, and Considerations for Greenhouse Gases*. Washington DC: Pew Center on Climate Change.

ENDS 1997a. Electricity industry sceptical on renewables, 25 March.

ENDS 1997b. Parliament demands renewable energy treaty, 16 May.

ENDS 1997c. Commission set to boost renewable energy, 21 November.

ENDS 1997d. Green groups welcome EU renewables target, 27 November.

ENDS 1998a. EU to harmonise renewables support schemes, 18 March.

ENDS 1998b. Energy ministers give their day to green issues, 12 May.

ENDS 1998c. MEPs call for national renewable energy goals, 22 June.

ENDS 1998d. Industry calls for EU climate agreements, 9 October.

ENDS 1998e. European quotas for renewable energy mooted. 10 December

ENDS 1999a. Plan for EU renewable energy law shelved, 9 February.

ENDS 1999b. Strong EU energy law demanded, 28 April.

ENDS 1999c. EU ministers request renewables framework, 12 May.

ENDS 1999d.Commission bids to salvage renewables directive, 3 December.

ENDS 2000a. European green power trading takes off, 22 February.

ENDS 2000b. MEPs demand strong renewables support, 31 March.

ENDS 2000c. EU renewable energy support directive proposed, 10 May.

ENDS 2000d. European building firms pledge greener future, 10 June.

ENDS 2000e. Insulation firms slam EU energy efficiency plan, 10 August.

ENDS 2000f. German renewables support 'not a state subsidy', 27 October.

ENDS 2000g. EU renewable law gets ministerial approval, 7 December.

ENDS 2001a. Germany to boost housing energy efficiency, 7 March.

ENDS 2001b. German green power law 'in the clear', 13 March.

ENDS 2001c. EU aims for more energy-efficient buildings, 26 April.

ENDS 2001d. Ministers neuter EU building efficiency drive, 5 December.

ENDS 2001e. MEP defy minister over buildings efficiency, 19 December.

ENDS 2002a. German buildings energy law enters into force, 1 February.

ENDS 2002b. Buildings energy law deadline not yet settled, 27 August.

ENDS 2002c. EU buildings efficiency law agreed, 3 October.

ENDS 2005a. 'Small firms good, big firms bad' says Trittin, 10 March.

ENDS 2005b. Europe eyes goal of large-scale carbon capture, 20 April.

ENDS 2005c. German renewable feed-in tariffs under attack, 3 May.

ENDS 2006. EU 'Must get real on energy-climate policies', 24 November.

ENDS 2007a. Green energy division looms over EU summit, 6 March.

ENDS 2007b. EU leaders pass climate commitment test, 9 March.

ENDS 2008a. EU building energy law review options outlined, 1 February.

ENDS 2008b. Analysis: prices and targets pressure shifts focus to lower demand, 1 February.

ENDS 2008c. Energy ministers welcome climate package, 28 February.

ENDS 2008d. Commission cautious on EU building energy limits, 28 April.

ENDS 2008e. MEP calls for 'double CO2 credits' to boost CCS, 6 May.

ENDS 2008f. Council set to reject EU-wide renewables trading, 27 June.

ENDS 2008g. Support grows for binding CCS obligation, 10 September.

ENDS 2008h. MEPs back beefed-up renewable energy law, 11 September.

ENDS 2008i. Concerns over upcoming building energy plans, 27 October.

ENDS 2008j. EU states find common ground on carbon storage, 30 October.

ENDS 2008k. Brussels wants to cut proposed funding for CCS, 12 November.

ENDS 2009a. EU building energy law revision 'must go further', 6 January.

ENDS 2009b. MEPs demand financing focus in buildings plan, 20 January.

ENDS 2009c. EU states doubts over greening building plans, 16 February.

ENDS 2009d. MEPs call for EU efficiency fund for buildings, 1 April.

ENDS 2009e. MEP set 2019 deadline for zero-energy buildings, 23 April.

ENDS 2009f. States' fears for green building law revision grow, 3 July.

ENDS 2009g. Ministers 'moving closer' to MEPs on green buildings, 24 July.

ENDS 2009h. Legislators to agree 'zero-buildings' deadline, 12 October.

ENDS 2009i. All new buildings to be 'near-zero-energy' by 2020, 18 November.

ENDS 2009j. Europe's political miscalculation. December 2009

ENDS 2010. Draft ETS benchmarks slipping beyond the summer, 14 June.

ENDS 2012. Poland still unhappy with the EU low-carbon roadmap, 7 March.

ENDS Report 2008. Call for free allowances to fund carbon storage, No. 400, 14 May.

ENEL 2009. *Annual Report 2008*. Rome: ENEL.

Energy Star 2011. *Major Milestones*. [Online]. Available at: www.energystar.gov/index.cfm?c=about.ab_milestones [accessed: 30 January 2012].

ENI 2009. CCS – *Carbon Dioxide (CO₂) Capture and Storage*. [Online]. Available at: www.eni.com/attachments/innovazione-tecnologia/technological-answers/scheda-cattura-sequestrazione-co2-eng.pdf [accessed: 9 February 2012].

EnR 2008. *Implementation of the EU Energy Performance of Buildings Directive – A Snapshot Report*. European Energy Network.

E.ON 2009 *Company Report 2008*. Düsseldorf: E.ON.

EPF 2011. *European Property Federation*. [Online]. Available at: www.epf-fepi.com/index2.html [accessed: 17 June 2011].

EREC 2004. *Campaign for Take-Off: Renewable Energy for Europe 1999–2003*. Brussels: European Renewable Energy Council.

EREC 2009. ITRE vote on the recast of the Energy Performance of Buildings Directive. Press release. 31 March.

Erhorn, H. and Erhorn-Klutting, H. 2009. *Germany: Impact, Compliance and Control of Legislation.* ASIEPi information paper P177. [Online]. Available at: www.buildup.eu/publications/7049 [accessed: 9 February 2012].

EU Energy 2007. Energy intensive users to get special treatment on CO_2 cuts. *EU Energy* 172–173, 14 December, 4–5.

EU Energy 2008. Poland rejects plans for 100 per cent ETS CO2 permit auctions. *EU Energy* 179, 21 March, 5.

EUFORES 2010. *II. Inter-Parliamentary Meeting. Madeira, Portugal 12–14 May 2000.* [Online]. Available at: www.eufores.org/index.php?id=46 [accessed: 9 February 2012].

EU Observer 2009. New committee chairs set to create bruising encounters. 16 July.

Euractiv 2007a. EU ministers endorse new energy policy for Europe. 29 June.

Euractiv 2007b. EU citizens air doubts about 'clean fossil fuels'. 21 September.

Euractiv 2007c. EU climate and energy proposals delayed. 23 October.

Euractiv 2008a. Interview: no 'huge' CO2 cuts for coal after 2020. 5 May.

Euractiv 2008b. Uncertainty over CO2 capture in 'fossil future'. 30 May.

Euractiv 2008c. Climate votes offer missing piece in Chinese coal puzzle. 8 October.

Euractiv 2008d. EU leaders clinch deal on CO2 storage financing. 12 December.

EURELECTRIC 1999. *Greenhouse Gas and Electricity Simulation, An Exercise in Trading Carried Out by the Electricity Industry in Collaboration with the International Energy Agency and Parisbourse,* October, Report 1999-420-0013.

EURELECTRIC 2000. *Gets 2 Report.* 6 November.

EURELECTRIC 2005. *Integration Electricity Markets through Wholesale Markets: EURELECTRIC Road Map to Pan-European Market.* Ref. 005-308-0010. Brussels: EURELECTRIC.

EURELECTRIC 2007. *Position paper – Review of the EU Emissions Trading Directive (2003/87/EC) and the Linking Directive (2004/101/EC).* July. Brussels: EURELECTRIC.

EURELECTRIC 2011. *National Renewable Energy Action Plans: An Industry Analysis.* A EURELECTRIC Renewables Action Plan (RESAP) Report. Brussels: EURELECTRIC.

Eurima 2011. *Our Members/Headquarters.* [Online]. Available at: www.eurima.org/headquarters/ [accessed: 9 February 2012].

EuroACE 2010. *Making Money Work for Buildings. Financial and Fiscal Instruments for Energy Efficiency in Buildings.* A report by Klinckenberg Consultants for EuroACE. Brussels: EuroACE.

EuroACE 2011. *EuroACE in a Nutshell.* [Online]. Available at: www.euroace.org/AboutUs/EuroACEinaNutshell.aspx [accessed: 17 June 2011].

EUROFER 2000. *The European Steel Industry and Climate Change.* Brussels: EUROFER.

EUROFER 2011. Steel industry goes to European Court on EU emissions trading scheme. Press release, Brussels, 21 July 21.

Europa 2010. Climate change: European Union notifies EU emission reduction targets following Copenhagen Accord. Brussels, 28 January. [Online]. Available at: http://europa.eu/rapid/pressReleasesAction.do?reference=IP/10/97 [accessed: 9 February 2012].

Europa 2011. Renewable energy targets: Commission calls on member states to boost cooperation. Brussels, 31 January. [Online]. Available at: http://europa.eu/rapid/pressReleasesAction.do?reference=IP/11/113 [accessed: 30 March 2012].

Europa 2012. Summaries of EU legislation. [Online]. Available at: http://europa.eu/legislation_summaries/index_en.htm [accessed: 30 March 2012].

European Commission 1975. *Community Program for Rational Use of Energy (RUE)*. Information. Energy.

European Commission 1979. *Third Report on the Community's Programme for Energy Saving.* Communication from the Commission to the Council. COM (79) 313 final.

European Commission 1984a. *Comparison of Energy Saving Programmes of EC Member States.* Communication from the Commission to the Council. COM (84) 36 final.

European Commission 1984b. *Towards a European Policy for the Rational Use of Energy in the Building Sector*. Communication from the Commission to the Council. COM (84) 614 final.

European Commission 1987a. *Towards a Continuing Policy for Energy Efficiency in the European Community*. Communication from the Commission. COM (87) 223 final.

European Commission. 1987b. Resolution of the Council of the European Communities and of the representatives of the Governments of the Member States, meeting with the Council of 19 October 1987 on the continuation and implementation of a European Community policy and action programme on the environment (1987–1992), *Official Journal of the European Communities* C 328, vol.30, 7 December.

European Commission 1988. *The Greenhouse Effect and the Community*. COM(88) 656, 16 November.

European Commission 1990. Proposal for a Council Decision Concerning the Promotion of Energy Efficiency in the Community. COM (90) 365 final.

European Commission 1991. *A Community Strategy to Limit Carbon Dioxide Emissions and to Improve Energy Efficiency*. SEC (91) 1744, 14 October.

European Commission 1992a. *Specific Actions for a Greater Penetration for Renewable Energy Sources: ALTENER.* COM (92) 180 final.

European Commission 1992b. *A Community Strategy to Limit Carbon Dioxide Emissions and to Improve Energy Efficiency*. COM (92) 246, 1 June.

European Commission 1992c. *Proposal for a Council Directive to Limit Carbon Dioxide Emissions by Improving Energy Efficiency (SAVE programme)*. Commission Communication (92) 182 final, 26 June.

European Commission 1993. *Amended Proposal for Council Decision Concerning the Promotion of Renewable Energies in the Community ALTERNER Programme.* COM (93) 278 Final.

European Commission 1996. *Energy for the Future: Renewable Sources of Energy.* Green paper for a Community strategy. Communication from the Commission, COM (96) 576 final.

European Commission 1997. *Energy for the Future: Renewable Sources of Energy.* White Paper for a Community Strategy and Action Plan. COM(97)599, 26 November.

European Commission 1998a. *Energy Efficiency in the European Community – Towards a Strategy for the Rational Use of Energy.* Communication from the Commission. COM (1998) 246 final.

European Commission 1998b. *Climate Change – Towards an EU Post-Kyoto Strategy.* Communication from the Commission to the Council and the European Parliament. COM (98)353, June.

European Commission 2000a. *Communication from the Commission to the Council and the European Parliament on EU Policies and Measures to Reduce Greenhouse Gas Emissions: Towards a European Climate Change Programme (ECCP).* COM(2000) 88 final, 8 March.

European Commission 2000b. *Green Paper on Greenhouse Gas Emissions Trading within the European Union.* COM(2000) 87 final, 8 March.

European Commission 2000c. *Green Paper: Towards a European Strategy for the Security of Energy Supply.* COM(2000) 769 final, 29 November.

European Commission 2000d. *Amended proposal for a Directive of the European Parliament and of the Council on the promotion of electricity from renewable energy sources in the internal electricity market.* COM(2000) 884 final, 29 December.

European Commission 2001a. *European Climate Change Programme, Final Report.* June.

European Commission 2001b. *Competitiveness of the Construction Industry. An Agenda for Sustainable Construction Practices in Europe. A report drawn up by the Working Group for Sustainable Construction with participants from the European Commission, Member States and Industry.* Brussels, 20 May.

European Commission 2001c. *Proposal for a Directive of the European Parliament and of the Council Establishing a Framework for Greenhouse Gas Emissions Trading within the European Community and Amending Council Directive 96/61/EC.* COM(2001) 581, 23 October.

European Commission 2005a. *The Support of Electricity from Renewable Energy Sources.* Communication from the Commission. COM 627, 7 December.

European Commission 2005b. *Reducing the Climate Change Impact of Aviation.* Communication from the Commission to the Council, the European Parliament, the European Economic and Social Committee and the Committee of the Regions. COM(2005) 459 final, 27 September.

European Commission 2005c. *Winning the Battle Against Global Climate Change*. Communication from the Commission to the Council, the European Parliament, the European Economic and Social Committee and the Committee of the Regions. COM(2005) 35 final, 9 February.

European Commission 2005d. *Doing More with Less*. Green Paper. COM(2005) 265 final.

European Commission 2006a. *Building a Global Carbon Market – Report Pursuant to Article 30 of Directive 2003/87/EC*. Communication from the Commission to the Council, the European Parliament, the European Economic and Social Committee and the Committee of the Regions. COM(2006)676 final, 13 November.

European Commission 2006b. Energy, environment, competiveness: Commission launches high level group. Press release IP/06/226, 24 February.

European Commission 2006c. *Green Paper – A European Strategy for Sustainable, Competitive and Secure Energy*. SEC(2006) 317, 8 March.

European Commission 2006d. *Action Plan for Energy Efficiency: Realising the Potential*. Communication from the Commission. COM (2006)545 final, 19 October.

European Commission 2006e. *Report of Working Group 3: Carbon Capture and Geological Storage (CCS), The Second European Climate Change Programme*. June.

European Commission 2007a. *Limiting Global Climate Change to 2 degrees Celsius. The Way Ahead for 2020 and Beyond*. Communication from the Commission. COM(2007) final.

European Commission 2007b. *Prospects for the Internal Gas and Electricity Market*. Communication from the Commission to the Council and the European Parliament. COM(2006) 841 final, 10 January.

European Commission 2007c. *Proposal for a Directive of the European Parliament and of the Council on the Promotion of the Use of Energy from Renewable Sources*. Version 6.3.3. Leaked 5 December.

European Commission 2007d. *Final Report of the 3rd Meeting of the ECCP Working Group on Emissions Trading on the Review of the EU ETS on Further Harmonisation and Increased Predictability*. 070521-22 Final report M3, 21 and 22 May.

European Commission 2007e. *An Energy Policy for Europe*. Communication from the Commission. COM(2007)1, 10 January.

European Commission 2007f. *Sustainable Power Generation from Fossil Fuels: Aiming for Near-zero Emissions from Coal after 2020*. Communication from the Commission to the Council and the European Parliament. COM(2006) 843 final, 10 January.

European Commission 2008a. *Attitudes of European Citizens Towards the Environment. Special Eurobarometer Report*. Brussels: European Commission.

European Commission 2008b. *20 20 by 2020 – Europe's Climate Change Opportunity*. Communication from the Commission. COM (2008) 30, 23 January.

European Commission 2008c. *Questions and Answers on the Revised EU Emissions Trading System.* Memo/08/796, 17 December.

European Commission 2008d. *Proposal for a Directive of the European Parliament and of the Council on the Promotion of the Use of Energy from Renewable Sources.* COM (2008) 30 final/19 final, SEC 2008 57, SEC 2008 85, 23 January.

European Commission 2008e. *Questions and Answers on the Directive on the Geological Storage of Carbon Dioxide.* Memo/08/798, 17 December.

European Commission 2008f. *Supporting Early Demonstration of Sustainable Power Generation from Fossil Fuels.* Communication from the Commission to the Council, the European Parliament, the European Economic and Social Committee and the Committee of the Regions. COM(2008)30 final, 23 January.

European Commission 2008g. *Proposal for a Directive of the European Parliament and the Council Amending Directive 2003/87/EC So as to Improve and Extend the Greenhouse Gas Emission Allowance Trading Scheme of the Community.* COM(2008) 16 final, 23 January.

European Commission 2008h. *Proposal for a Directive of the European Parliament and of the Council on the Energy Performance of Buildings* (recast) (presented by the Commission). COM(2008) 780 final 2008/0223 (COD), 13 November.

European Commission 2008i. *Accompanying Document to the Proposal for a Recast of the Energy Performance of Buildings Directive (2002/91/EC).* Summary of the impact assessment. Commission staff working document. 13-XI-2008 SEC(2008) 2865.

European Commission 2008j *The Support of Electricity from Renewable Energy Sources.* Commission staff working document, SEC 57. Brussels: European Commission.

European Commission 2008k. *Impact Assessment. Document Accompanying the Package of Implementation measures for the EU's Objectives on Climate Change and Renewable Energy for 2020.* Commission staff working document. SEC(2008) 85/3

European Commission. 2008 l. *Impact assessment, Accompanying Document to the Proposal for a Directive of the European Parliament and the Council amending Directive 2003/87/EC so as to improve and extend the EU greenhouse gas emission allowance trading system.* Commission Staff Document, COM(2008) 16 final, 23 January.

European Commission 2008m. *Commission Staff Working Document, Annex to the Impact Assessment,* document accompanying the package of implementation measures for the EU's objectives on climate change and renewable energy for 2020, SEC(2008) 85, 27 February.

European Commission 2008n. *Proposal for a Directive of the European Parliament and of the Council on the geological storage of carbon dioxide and amending Council Directives 85/337/EEC, 96/61/EC, Directives 2000/60/EC, 2001/80/EC, 2004/35/EC, 2006/12/EC and Regulation (EC) No 1013/2006,* COM(2008) final, 23 January.

European Commission 2008o. Eurobarometer 68 – Public opinion in the European Union. Brussels: European Commission.

European Commission 2009a. Emissions trading: member states approve list of sectors deemed to be exposed to carbon leakage. Press release IP/09/1338. 18 September.

European Commission 2009b. *Commission Decision of 30 June 2009 Establishing a Template for National Renewable Energy Action Plans under Directive 2009/28/EC of the European Parliament and of the Council* (notified under document number C(2009) 5174). (2009/548/EC).

European Commission 2009c. Low energy buildings in Europe: current state of play, definitions and best practice. Commission info-note on low energy buildings. 25 September.

European Commission 2010a. *Analysis of Options to Move beyond 20% Greenhouse Gas Emission Reductions and Assessing the Risk of Carbon Leakage*. COM(2010)265, 26 May.

European Commission 2010b. *EU Energy and Transport in Figures*. Statistical pocketbook. Luxembourg: Publications Office of the European Union.

European Commission 2010c. Commission Decision of 3 November 2010 Laying Down Criteria and Measures for the Financing of Commercial Demonstration Projects of Innovative Renewable Energy Technologies under the Scheme for Greenhouse Gas Emissions Allowance Trading within the Community Established by Directive 2003/87/EC of the European Parliament and the Council. *Official Journal of the European Union* L290/39, 6 November.

European Commission 2011a. Renewable energy targets: Commission calls on member states to boost cooperation. Press release, 31 January.

European Commission 2011b. Meeting document for the Expert Workshop on the comparative framework methodology for cost optimal minimum energy performance requirements. In preparation of a delegated act in accordance with Art 290 TF EU 6 May 2011 in Brussels, presented by the Directorate General for Energy.

European Commission 2011c. *Energy Efficiency Plan 2011*. Communication from the Commission to the European Parliament, the Council, the European Economic and Social Committee and the Committee of the Regions. COM(2011) 109 final, 8 March.

European Commission 2011d. Commission Decision of 27 April 2011 Determining Transitional Union-wide Rules for Harmonised Free Allocation of Emission Allowances Pursuant to Article 10a of Directive 2003/87/EC. Decision 2011/278/EU. *Official Journal of the European Union*, L 130/1, 17 May.

European Commission 2011e. Special Eurobarometer 372 – Climate change. Brussels: European Commission.

European Commission 2011f. *A Roadmap for Moving to a Competitive Low Carbon Economy in 2050*. Communication from the Commission to the European Parliament, the Council, the European Economic and Social Committee and the Committee of the Regions, COM(2011) 112 Final, 8 March.

European Council 1993. Directive 93/76/EEC of 13 September 1993 to Limit Carbon Dioxide Emissions by Improving Energy Efficiency (SAVE).

European Council 2004. *Presidency Conclusions.* Document 9048/04. 25 and 26 March.

European Council 2005. *Presidency Conclusions.* Document 7619/1/05. 22 and 23 March.

European Council 2006. *Presidency Conclusions.* Document 7775/06. 23 and 24 March.

European Council 2007. *Presidency Conclusions.* Document 7224/1/07. 8 and 9 March.

European Council 2009a. Decision No 406/2009/EC of the European Parliament and the Council of 23 April 2009 on the effort of Member States to reduce their greenhouse gas emissions to meet the Community's greenhouse gas emission reduction commitments up to 2020, *Official Journal* L 140/136, 6 June.

European Council 2009b. *Note from Secretary General to Delegations.* Subject: Energy efficiency package. Council of the European Union. Interinstitutional File: 2008/0222 (COD) 2008/0221 (COD) 2008/0223 (COD). 29 May.

European Energy Review 2011. CCS facing mounting obstacles, 10 February.

European Environment Agency 2004. *Greenhouse Gas Emissions Trends and Projections in Europe 2004.* EEA Report No. 5/2004. Copenhagen: European Environment Agency.

European Parliament 1986. Resolution on the measures to counteract the rising concentration of carbon dioxide in the atmosphere (the 'Greenhouse' effect). *OJ* C255, 13 October.

European Parliament 2000. *I Report on the Proposal for a European Parliament and Council Directive on the Promotion of Electricity from Eenewable Energy Sources in the Internal Electricity Market* (COM(2000) 279 – C5-0281/2000 – 2000/0116(COD)). 30 October 2000, final. A5-0320/2000.

European Parliament 2001. *II Recommendation for Second Reading on the Proposal for a European Parliament and Council Directive on the Pomotion of Electricity from Renewable Energy Sources in the Internal Electricity Market* (COM(2000) 279 – C5-0281/2000 – 2000/0116(COD)). 22 June, final. A5-0227/2001.

European Parliament 2006. *Report with Recommendations to the Commission on Heating and Cooling from Renewable Sources of Energy* (2005/2122(INI)). 1 February 2006. Final. A6-0020/2006.

European Parliament 2008a. *Compromise Amendments 1-25 on the Proposal for a Directive of the European Parliament and the Council Amending Directive 2003/87/EC.* Draft report Avril Doyle, 2008/0013(COD), 5 October.

European Parliament 2008b. *Draft Report on the Proposal for a Directive of the European Parliament and of the Council on the Promotion of the Use of Energy from Renewable Sources* (COM(2008)0019 – C6-0046/2008 – 2008/0016(COD)) Committee on Industry, Research and Energy. 2008/0016 (COD), 13 May.

European Parliament 2008c. *Report on an Action Plan for Energy Efficiency: Realising the Potential* (2007/2106(INI)). A6-0003/2008.

European Parliament 2009a. *Report on the Proposal for a Directive of the European Parliament and of the Council on the Energy Performance of Buildings* (recast) (COM(2008)0780 – C6-0413/2008 – 2008/0223(COD)).

European Parliament 2009b. *Activity Report of the Committee on the Environment, Public Health and Food Safety 2004–2009 Parliament.* Brussels: European Parliament.

European Parliament and Council 2001. Directive 2001/77/EC on the Promotion of Electricity Produced from Renewable Energy Sources in the Internal Electricity Market. Brussels: European Commission.

European Parliament and Council 2002. Directive 2002/91/EC of the European Parliament and of the Council of 16 December 2002 on the energy performance of buildings. *Official Journal of the European Union*, L1/65, 4 January.

European Parliament and Council 2003. Directive 2003/87/EC of the European Parliament and of the Council of 13 October 2003 Establishing a Scheme for Greenhouse Gas Emission Allowance Trading Within the Community and Amending Council Directive 96/61/EU. *Official Journal of the European Union*, L 275, 25 October.

European Parliament and Council 2004. Directive 2004/101/EC of the European Parliament and of the Council of 27 October 2004 amending Directive 2003/87/EC Establishing a Scheme for Greenhouse Gas Emission Allowance Trading Within the Community, in Respect of the Kyoto Protocol's Project Mechanisms. *Official Journal of the European Union*, L 338/18, 13 November.

European Parliament and Council 2005. Directive 2005/32/EC of the European Parliament and of the Council of 6 July 2005 establishing a framework for the setting of ecodesign requirements for energy-using products and amending Council Directive 92/42/EEC and Directives 96/57/EC and 2000/55/EC of the European Parliament and of the Council.

European Parliament and Council 2009a. Directive 2009/28/EC of the European Parliament and of the Council of 23 April 2009, on the promotion of the use of energy from renewable sources and amending and subsequently repealing Directives 2001/77/EC and 2003/30/EC. *Official Journal of the European Union*, L 140/16, 5 June.

European Parliament and Council 2009b. Directive 2009/29/EC of the European Parliament and the Council of 23 April 2009 amending Directive 2003/87/EC so as to improve and extend the greenhouse gas emission allowance trading scheme of the Community. *Official Journal of the European Union*, L140/63, 5 June.

European Parliament and Council 2009c. Directive 2009/31/EC of the European Parliament and of the Council of 23 April 2009 on the geological storage of carbon dioxide and amending Council Directive 86/337/EEC, European Parliament and Council Directives 2000/60/EC, 2001/80/EC, 2004/35/EC,

2006/12/EC, 2008/1/EC and Regulation (EC) No 1013/2006. *Official Journal of the European Union*, L140/114, 5 June.

European Parliament and Council 2010. Directive 2010/317/EU of the European Parliament and of the Council of May 2010 on the energy performance of buildings (recast). *Official Journal of the European Union*, L153/13, 18 June.

Fairbrass, J. and Jordan, A. 2004. Multi-level governance and environmental policy, in *Multi-level Governance*, edited by I. Bache and M. Flinders. Oxford: Oxford University Press, 147–64.

FIEC 2010. *Key Figures. Activity 2009*. Brussels: FIEC.

Financial Times 2008a. Europe 500 2008. [Online]. Available at: www.ft.com/cms/s/0/05b22652-413c-11dd-9661-0000779fd2ac,dwp_uuid=4ca4e366-413b-11dd-9661-0000779fd2ac.html?nclick_check=1 [published 24 June 2008].

Financial Times 2008b. France poaches top climate change eurocrat. [Online]. Available at: www.ft.com/intl/cms/s/0/95854af0-38d3-11dd-8aed-0000779fd2ac.html#axzz1RnhfJ8Fn [accessed: 11 July 2011].

Fioretos, O. 2011. Historical institutionalism in international relations. *International Organization*, 65 (Spring), 367–99.

Finnemore, M. and K. Sikkink. 1998. International norm dynamics and political change. *International Organization*, 52, 887–917.

Flåm, K.H. 2009. Restricting the import of 'emission credits' in the EU: a power struggle between states and institutions. *International Environmental Agreements*, 9(1), 23–38.

Fligstein, N. 1997. Social skill and institutional theory. *American Behavioral Scientist*, 40(4), 397–405.

Fligstein, N. 2001. *The Architecture of Markets*. Princeton, NJ: Princeton University Press.

Fligstein, N. 2008. *Euroclash*. Oxford: Oxford University Press.

Fligstein, N. and Mara-Drita, I. 1996. How to make a market: reflections on the attempt to create a Single Market in the European Union. *American Journal of Sociology*, 102 (1), 1–33.

Fligstein. N and Stone Sweet, A. 2002. Constructing polities and markets: an institutionalist account of European integration. *The American Journal of Sociology*, 107(5), 1206–43.

Foquet, D and Johansson, T.B. 2008. European renewable energy policy at crossroads: focus on electricity support mechanisms. *Energy Policy*, 36(11), 4079–92.

Friedland, R. and Alford, R.R. 1991. Bringing society back in: symbols, practices, and institutional contradictions, in *The New Institutionalism in Organizational Analysis*, edited by W.W. Powell and P. DiMaggio. Chicago, IL: The University of Chicago Press, 232–63.

Friends of Europe 2008. *Carbon Capture and Storage – Making It Happen*. Report from a meeting held in Brussels, May 2008. Brussels: Friends of Europe, Bellona.

Fuglseth, B.B. 2009. *Regulative Change Targeting Energy Performance of Buildings in Sweden: Key Drivers and Main Implications*. FNI Report 2/2009. Lysaker: Fridtjof Nansen Institute.

G8 2008. *Joint Statement by G8 Energy Ministers*. Aomori, Japan, on 8 June 2008.

GDF Suez 2009. 2008 Reference document. Paris: DGF Suez.

Geller, H., Harrington, P., Rosenfeld, A.H., Tanishima, S. and Unander, F. 2005. Polices for increasing energy efficiency: thirty years of experience in OECD countries. *Energy Policy*, 34(15), 556–73.

George, A.L. and Bennett, A. 2005. *Case Studies and Theory Development in the Social Sciences*. Cambridge, MA: MIT Press.

Gerring, J. 2008. Case selection for case-study analysis: qualitative and quantitative techniques', in *The Oxford Handbook of Political Methodology*, edited by J.M. Box-Steffensmeier, H.E. Brady and D. Collier. Oxford: Oxford University Press, 645–84.

Glachant, J.M. 2003. The making of competitive electricity markets in Europe: no single way and no single market, in *Competition in European Electricity*, edited by D. Finon and J.M. Glachant. Cheltenham: Edward Elgar, 7–38.

Glass for Europe 2011. Member companies of Glass for Europe. [Online]. Available at: www.glassforeurope.com/en/about/our-members.php [accessed: 17 June 2011].

Grant, W. 2011. Business: The elephant in the room? in *Environmental Policy-making in Britain, Germany and the European Union*, edited by R.K.W. Wurzel. Manchester: Manchester University Press, 198–213.

Grant, W., Matthews, D. and Newell, P. 2000. *The Effectiveness of European Union Environmental Policy*. London: Macmillan Press.

Greenpeace 2008. *False Hope – Why Carbon Capture and Storage Won't Save the Climate*. Amsterdam: Greenpeace.

Grubb, M., Azar, C. and Persson, M. 2005. Allowance allocation in the European emissions trading system: a commentary. *Climate Policy*, 5, 127–36.

Grubb, M. and Gupta, J. 2000. Implementing European leadership, in *Climate Change and European Leadership*, edited by M. Grubb and J. Gupta. Dordrecht: Kluwer Academic Publishers, 287–313.

Grubb, M., Vrolijk, C. and Brack, D. 1999. *The Kyoto Protocol: A Guide and Assessment*. London: Earthscan.

Gullberg, A.T. 2008. Lobbying friends and foes in climate policy: the case of business and environmental interest groups in the European Union. *Energy Policy*, 36(8), 2964–72.

Haas, E. [1958] 2004. *The Uniting of Europe: Politics, Social and Economic Forces, 1950–1957*. 3rd edition. Notre Dame, IN: University of Notre Dame Press.

Haas, E. 1970. The study of regional integration: reflections on the joy and anguish of pretheorizing. *International Organization*, 24(4), 607–46.

Haigh, N. 1996. Climate change policies and politics in the European Community, in *Politics of Climate Change*, edited by T. O'Riordan and J. Jager. London: Routledge, 155–86.

Haigh, N. (ed.) 2011. *Manual of Environmental Policy – the EU and Britain.* London: Institute for European Environmental Policy.

Hall, P.A. 2003. Aligning ontology and methodology in comparative research, in *Comparative Historical Analysis in the Social Sciences,* edited by J. Mahoney and D. Rueschemeyer. Cambridge: Cambridge University Press, 373–444.

Hall, P.A. and D. Soskice. 2001. Introduction, in *Varieties of Capitalism: The Institutional Foundation of Comparative Advantage*, edited by P.A. Hall and D. Soskice. Oxford: Oxford University Press, 1–69.

Hampton, K. 2008. Mail correspondence between European Parliamentarian Chris Davies and head of Policy at Climate Change Capital, Kate Hampton, 13 November 2008. [Online]. Available at: www.corporateeurope.org/system/files/../20081111-Davies-Hampton.pdf [accessed: 10 July 2011].

Hardy, C. and S. Maguire. 2008. Institutional entrepreneurship, in *The SAGE Handbook of Organizational Institutionalism*, edited by R. Greenwood, C. Oliver, K. Sahlin and R. Suddaby Thousand Oaks, CA: Sage, 198–217.

Hasselmeier, G. and Wettestad, J. 2000. *German Climate Policy Ambitiousness: Just a Side Effect of Reunification?* FNI Report 2/2000. Lysaker: Fridtjof Nansen Institute.

Hatch, M.T. 2007. The politics of climate change in Germany: domestic sources of environmental foreign policy, in *Europe and Global Climate Change. Politics, Foreign Policy and Regional Cooperation*, edited by P. Harris. Cheltenham: Edward Elgar, 41–62.

Hedström, P. 2008. Studying mechanisms to strengthen causal inferences in quantitative research', in *The Oxford Handbook of Political Methodology*, edited by J.M. Box-Steffensmeier, H.E. Brady and D. Collier. Oxford: Oxford University Press, 319–35.

Helm, D. 2009. EU climate policy: a critique, in *The Economics and Politics of Climate Change*, edited by D. Helm and C. Hepburn. Oxford: Oxford University Press, 222–62.

High Level Group on Competitiveness, Energy and the Environment 2007. *Fifth Report – Addressing both International Action on Climate Change and Better Regulation.* Brussels. 8 November [Online]. Available at: http://ec.europa.eu/enterprise/policies/sustainable-business/files/environment/hlg/doc_07/hlg-fifth-08-11-07_en.pdf [accessed: 10 February 2012].

Hildingsson, R., Stripple, J. and Jordan, A. 2010. Renewable energies: a continuing balancing act? in *Climate Change Policy in the European Union: Confronting Dilemmas of Mitigation and Adaptation*, edited by A. Jordan, D. Huitema, H. van Asselt, T. Rayner and F. Berkhout. Cambridge: Cambridge University Press, 103–24.

Hix, S. 2005. *The Political System of the European Union*. Hampshire: Palgrave Macmillian.

HM Government 2007. *Meeting the Energy Challenge: A White Paper on Energy*. London: HM Government.

HM Government 2009. *The UK Renewable Energy Strategy*. London: HM Government. [Online]. Available at: www.official-documents.gov.uk/document/cm76/7686/7686.pdf [accessed: 10 February 2012].

Hooghe, L. 2001. *The European Commission and the Integration of Europe. Images of Governance*. Cambridge: Cambridge University Press.

Hooghe, L. and Marks, G. 2001. *Multi-Level Governance and European Integration*. Oxford: Rowman & Littlefield.

Huitema, D. and S.Meijerink. 2010. Realizing water transitions. The role of policy entrepreneurs in water policy change. *Ecology and Society* 15(2): 26. Online URL: http://www.ecologyandsociety.org/vol15/iss2/art26/ (SENSE B classification).

Iberdrola. *Iberdrola: A History with Deep Roots*. [Online]. Available at: www.iberdrola.es/webibd/ngc/en/micro/historia/index.htm. [accessed: 10 February 2012].

Ikwue, J. and J. Skea. 1994. Business and the genesis of the European Community Carbon Tax Proposal, *Business Strategy and the Environment*, 3(2), 1–11.

International Environment Reporter. 2007. EU industry chief calls for measures to ensure competitiveness amid trading, 30(25), 12 December, 986–98.

International Environment Reporter 2008. EU lawmakers back emissions limits for power plants; seek more carbon storage, 31(21), 15 October.

Isover 2011. Saint-Gobain Isover a/s. [Online]. Available at: www.isover.dk/ [accessed: 10 February 2012].

Jachtenfuchs, M. 2001. The governance approach to European integration. *Journal of Common Market Studies*, 39(2), 245–64.

Jacobsson, S. and Bergek, A. 2004. Transforming the energy sector: the envolution of technological systems in renewable energy technology. *Industrial and Corporate Change*, 13(5), 815–47.

Jacobsson, S. and Lauber, V. 2006. The politics and policy of energy system transformation – explaining the German diffusion of renewable energy technology. *Energy Policy*, 34(3), 256–76.

Jäger, J. and O'Riordan, T. 1996. The history of climate change science and politics, in *Politics of Climate Change*, edited by T. O'Riordan and J. Jäger. London: Routledge, 1–32.

Jänicke, M. 2011. German climate change policy: political and economic leadership, in *The European Union as a Leader in International Climate Change Politics*, edited by R.K.W. Wurzel and J. Connelly. Oxford: Routledge, 129–47.

Jankowska, K. 2011. Poland's climate change policy struggle: greening the East, in *The European Union as a Leader in International Climate Change Politics*, edited by R.K.W. Wurzel and J. Connelly. Oxford: Routledge, 163–79.

Jensen. O.M., K.B. Wittchen and K.E. Thomsen. 2009. *Towards Very Low Energy Buildings*. Ålborg, Denmark: Danish Building Research Institute.

Jilek, W. 2010. *Implementation of EPBD in Austria.* Concerted Action. Implementation of the Energy Performance of Buildings, updated Country Report for Austria. [Online]. Available at: www.buildup.eu/publications/16210 [accessed: 18 June 2011].

Jordan, A. 2001. The European Union: an evolving system of multi-level governance...or government? *Policy & Politics*, 29(2), 43–70.

Jordan, A. and Rayner, T. 2010. The evolution of climate policy in the European Union: an historical overview, in *Climate Change Policy in the European Union. Confronting the Dilemmas of Mitigation and Adaptation*, edited by A. Jordan, D. Huitema, H.V. Asselt, T. Rayner and F. Berkhout. Cambridge: Cambridge University Press, 52–81.

Jordan, A., Huitema, D., Asselt, H.V., Rayner, T. and Berkhout, F. (eds) 2010. *Climate Change Policy in the European Union. Confronting the Dilemmas of Mitigation and Adaptation.* Cambridge: Cambridge University Press.

Jordan, A., Wurzel, R.K.W., Zito, A.R. and Brückner, L. 2003. Policy innovation or 'muddling through'? 'New' environmental policy instruments in the United Kingdom. *Environmental Politics*, 12(1), 179–98.

Jørgensen, K.E. 2006. Overview: The European Union and the world, in *Handbook of European Union Politics*, edited by K.E. Jørgensen, M.A. Pollack and B. Rosamond. Thousand Oaks, CA: Sage, 507–25.

Kemeny, J. 2001. Comparative housing and welfare. *Journal of Housing and the Built Environment*, 16: 53–70.

Keulenaer, H. and Gerwen, R. 2006. *The Passive House in the Electricity System of the Future,* short paper, Passive House Symposium 2006, Heusden-Zolder, Belgium, 6 October 2006.

Key Stakeholders Alliance for ETS Review 2007. *Lowering Production is No Benefit for the Environment, says European Industry.* Brussels: CEFIC, CEMBUREAU, CEPI, CERAMIE-UNIE, CPIV, EULA, EUROCHLOR, EUROFER, EUROMETAUX, IFIEC. [Online]. Available at: www.usgbv. com/userfiles/swf/Publicatie17.pdf [accessed: 10 February 2012].

Kingdon, J.W. [1984] 2011. *Agendas, Alternatives, and Public Policies.* Boston, MA: Little, Brown.

Knill, C. 2005. Cross-national policy convergence: causes, concepts and empirical findings. *Journal of European Public Policy*, 12(5), 775–96.

Kohler-Koch, B. and Rittberger, B. 2006. Review article: the 'governance turn' in EU studies. *Journal of Common Market Studies*, 44 (Annual Review), 27–47.

Krasner, S.D. 1983. Structural causes and regime consequences: regimes as intervening variables, in *International Regimes*, edited by S.D. Krasner. Ithaca, NY: Cornell University Press, 1–22.

Kruger, J. and Pizer, W.A. 2004. *The EU Emissions Trading Directive: Opportunities and Potential Pitfalls.* Discussion Paper RFF DP 04-24, Washington, DC: Resources for the Future.

Kuhn, T. 2001. Implications of the 'Preussen Elektra'. *Legal Issues of Economic Integration*, 28(3), 361–76.

Lacasta, N., Oberthur, S., Santos, E. and Barata, P. 2010. From sharing the burden to sharing the effort: Decision 406/2009/EC on Member State Emission Targets for non-ETS sectors, in *The New Climate Policies of the European Union*, edited by S. Oberthur and M. Pallemaerts. Brussels: VUB Press, 93–117.

Lawrence, T.B. and Suddaby, R. 2006. Institutions and institutional work, in *Handbook of Organization Studies*, 2nd Edition, edited by S.R. Clegg, C. Hardy, T.B. Lawrence, and W.R. Nord. London: Sage, 215–54.

Leca, B., J. Battilana and E. Boxenbaum. 2006. Taking stock of institutional entrepreneurship: What do we know? Where do we go?' Paper presented at the Academy of Management Meetings, 2006, Atlanta, GA.

Lenschow, A. 2006. Environmental policy in the European Union: bridging policy, politics and polity dimensions, in *Handbook of European Union Politics*, edited by K.E. Jørgensen, M.A. Pollack and B. Rosamond. London: Sage, 413–31.

Lindberg, L.N. 1963. *The Political Dynamics of European Economic Integration*. Stanford, CA: Stanford University Press.

Lounsbury, M. 2007. A tale of two cities. *Academy of Management Journal*, 50(2): 289–307.

Lov om fremme av vedvarende energi 2008. LOV nr 1392 af 27/12/2008. ([Norwegian] Law No. 1392 of 27 December 2008).

Macrory, R. and Hession, M. 1996. The European Community and climate change: the role of law and legal competence, in *Politics of Climate Change*, edited by T. O'Riordan and J.Jager. London: Routledge, 106–55.

Mahoney, J. 2003. Strategies of causal assessment in comparative historical analysis, in *Comparative Historical Analysis in the Social Sciences*, edited by J. Mahoney and D. Rueschemeyer. Cambridge: Cambridge University Press, 337–72.

Mahoney, J. and Goertz, G. 2006. A tale of cultures: contrasting quantitative and qualitative research. *Political Analysis*, 14, 227–49.

Mahoney, J. and Rueschemeyer, D. (eds). 2003. *Comparative Historical Analysis in the Social Sciences*. Cambridge: Cambridge University Press.

March, J.G. and Olsen, J.P. 1989. *Rediscovering Institutions*. New York: The Free Press.

March, J.G. and Olsen, J.P. 1998. The institutional dynamics of international political orders. *International Organization*, 52(4), 943–69.

Marks, G and Hooghe, L. 2004. Contrasting visions of multi-level governance, in *Multi-level Governance*, edited by I. Bache and F. Flinders. Oxford: Oxford University Press, 15–30.

Marks, G., Hooghe, L. and Blank, K. 1996. European integration from the 1980s: state-centric v. multi-level governance. *Journal of Common Market Studies*, 34(3), 341–78.

Mazey, S. and Richardson, J. 2006. Interest groups and EU policy making, in *European Union: Power and Policy Making*, edited by J. Richardson. Oxford: Routledge, 247–68.

McCormick, J. 1999. *Understanding the European Union*. London: Palgrave.

McKibben, H.E. 2010. Issue characteristics, issue linkage, and states' choice of bargaining strategies in the European Union. *Journal of European Public Policy*, 17(5), 694–707.

Meadowcroft, J. and Langhelle, O. (eds) 2009a. *Caching the Carbon – The Politics and Policy of Carbon Capture and Storage*. Cheltenham: Edward Elgar.

Meadowcroft, J. and Langhelle, O. 2009b. The politics and policy of carbon capture and storage, in *Caching the Carbon – The Politics and Policy of Carbon Capture and Storage*, edited by J. Meadowcroft and O. Langhelle. Cheltenham: Edward Elgar, 1–22.

Meijerink, S. and Huitema, D. 2010. Policy entrepreneurs and change strategies: lessons from sixteen case studies of water transitions around the globe. *Ecology and Society* 15(2): 21 Online URL: http://www.ecologyandsociety.org/vol15/iss2/art21/ (SENSE B classification).

Metz, B., Davidson, O., de Coninck, H., Loos, M. and Meyer, L. (eds) 2005. *Carbon Dioxide Capture and Storage, Special Report of the Intergovernmental Panel on Climate Change*. Cambridge: Cambridge University Press.

Metz, B., Davidson, O.R., Bosch, P.R., Dave, R. and Meyer, L.A. (eds) 2007. *Climate Change 2007: Mitigation*. Cambridge: Cambridge University Press.

Meyer, J.W. 2000. Globalization: sources and effects on nation states and societies. *International Sociology*, 15(2), 233–48.

Meyer, J.W., Frank, D.J., Hironka, A., Schofer, E. and Tuma, N.B. 1997. The structuring of a world environmental regime, 1870–1990. *International Organization*, 51(4), 623–51.

Meyer, N.I. 2003. European schemes for promoting renewables in liberalised markets. *Energy Policy*, 31, 665–76.

Meyer, N.I. 2004. Renewable energy policy in Denmark. *Energy for Sustainable Development*, 8(1), 25–35.

Meyer, J.W. and Rowan, B. 1991. Institutionalized organizations: formal structure as myth and ceremony, in *The New Institutionalism in Organizational Analysis*, edited by P.J. DiMaggio and W.W. Powell. Chicago, IL: University of Chicago Press, 41–62 [first published in *American Journal of Sociology* 1983].

Meyer, J.W., Boli, J., Thomas, G.M. and Ramirez, F.O. 1997. World society and the nation-state. *American Journal of Sociology*, 103 (1): 144–81.

Mintrom, M. 1997. Policy entrepreneurs and the diffusion of innovation. *American Journal of Political Science*, 41(3): 738–70.

Mitchell, C. and Connor, P. 2004. Renewable energy policy in the UK 1990–2003. *Energy Policy*, 32(17), 1935–47.

Molina, J.L. and Álvarez, S. 2009. *Spain: Impact, Compliance and Control of Legislation*. ASIEPI information paper P172. [Online]. Available at: www.buildup.eu/publications/7050 [accessed: 13 February 2012].

Moravcsik, A. 1993. Preferences and power in the European Community. A liberal intergovernmentalism approach. *Journal of Common Market Studies*, 31(4), 473–524.

Moravcsik. A. 1998. *The Choice for Europe*. Ithaca, NY: Cornell University Press.

Moravcsik, A. 2008. The New Liberalism, in *The Oxford Handbook of International Relations*, edited by C. Reus-Smit and D. Snidal. Oxford: Oxford University Press, 234–54.

Moravcsik. A. and Schimmelfenning, F. 2009. Liberal intergovernmentalism, in *European Integration Theory*, edited by T. Diez and A. Wiener. Oxford: Oxford University Press, 67–87.

NER 300. 2012. Website information on NER 300 [Online]. Available at: www.ner300.com/ [accessed: 30 March 2012].

Newell, P. 2000. *Climate for Change: Non-State Actors and the Global Politics of the Greenhouse*. Cambridge: Cambridge University Press.

Niemann, A. 2006. *Explaining Decisions in the European Union*. Cambridge: Cambridge University Press.

Niemann, A. and Schmitter, P.C. 2006. Neofunctionalism, in *Handbook of European Union Politics*, edited by K.E. Jørgensen, M.A. Pollack and B. Rosamond. Thousand Oaks, CA: Sage, 45–66.

Non-paper 2007. Joint proposal by Germany, Poland and the United Kingom on an alternative renewable flexibility mechanism. [Online]. Available at: www.endsreport.com/docs/20080723a.doc [accessed: 13 February 2012].

Nordqvist, J., Boyd, C. and Klee, H. 2002. Three big Cs: climate, cement and china. *Greener Management International*, 39(Autumn), 69–82.

Oberthür, S., 2006. The Climate Change Regime: Interactions with ICAO, IMO and the EU Burden Sharing Agreement, in *Institutional Interaction in Global Environmental Governance. Synergy and Conflict among International and EU Policies*, edited by S. Oberthür and T. Gehring. Cambridge, MA: MIT Press, 53–78.

Oberthür, S. and Dupont, C. 2011. The Council, the European Council and international climate policy, in *The European Union as a Leader in International Climate Change Politics*, edited by R.K.W. Wurzel and J. Connelly. Oxford: Routledge, 74–91.

Oberthür, S. and Gehring, T. 2006. Introduction, in *Institutional Interaction in Global Environmental Governance. Synergy and Conflict among International and EU Policies*, edited by S. Oberthür and T. Gehring. Cambridge, MA: MIT Press, 1–18.

Oberthür, S. and Gehring, T. 2011. Institutional interaction: ten years of scholarly development, in *Managing Institutional Complexity: Regime Interplay and Global Environmental Change*, edited by S. Oberthür and O.S. Stokke. Cambridge, MA: MIT Press, 25–58.

Oberthür, S. and Ott, H. 1999. *The Kyoto Protocol: International Climate Policy for the 21st Century*. Berlin: Springer.

Oberthür, S. and Pallemaerts, M. 2010. The EU's internal and external climate policies: an historical overview, in *The New Climate Policies of the European Union*, edited by S. Oberthur and M. Pallemaerts. Brussels: VUB Press, 27–65.

Oberthür, S. and Stokke, O.S. 2011. *Managing Institutional Complexity: Regime Interplay and Global Environmental Change*. Cambridge, MA: MIT Press.

Oberthür, S. and Tänzler, D. 2007. Climate policy in the EU: international regimes and policy diffusion, in *Europe and Global Climate Change – Politics, Foreign Policy and Regional Cooperation*, edited by P. Harris. Aldershot: Edward Elgar, 255–79.

OECD 2006. *Linking GHG Emission Trading Schemes and Markets*. Paris:OECD. [Online]. Available at: www.oecd.org/dataoecd/45/35/37672298.pdf [accessed: 13 February 2012].

Olsen, J.P. 2007. *Europe in Search of Political Order. An Institutional Perspective on Unity/Diversity, Citizens/Their Helpers, Democratic Design/Historical Drift and Co-existence of Orders*. Oxford: Oxford University Press.

O'Riordan, T. and Rowbotham, E. 1996. Struggling for credibility. The United Kingdom's response, in *Politics of Climate Change*, edited by T. O'Riordan and J. Jäger. London: Routledge, 228–68.

OSPAR 2007a. *OSPAR Decision 2007/1 to Prohibit the Storage of Carbon Dioxide Streams in the Water Column or on the Sea-bed, Annex 5*, adopted at the OSPAR meeting in Oostend, 25–29 June 2007.

OSPAR 2007b. OSPAR Decision 2007/2 on the Storage of Carbon Dioxide Streams in Geological Formations, Annex 6, adopted at the OSPAR meeting in Oostend, 25–29 June 2007.

PAI. 2006. *Poland's Fuel and Energy Sectors*. Warsaw: Polish Information and Foreign Investment Agency.

Panek, A. and Popiolek, M. 2009. *Poland: Impact, Compliance, and Control of Legislation*. ASIEPI information paper P171. [Online]. Available at: www.buildup.eu/publications/7045 [accessed: 13 February 2012].

Papadopoulou, K., Papaglastra, M., Laskari, M. and Santamouris, M. 2009. *Evaluation of the Impact of National EPBD Implementation in MS*. ASIEPI information paper P180. [Online]. Available at: www.buildup.eu/publications/7368 [accessed: 13 February 2012].

Peters, G. and Pierre, J. 2009. Governance approaches, in *European Integration Theory*, edited by A. Wiener and T. Diez. Oxford: Oxford University Press, 91–104.

Peterson, J. 1995. Decision-making in the European Union: towards a framework for analysis. *Journal of European Public Policy*, 2(1), 69–93.

Peterson. J. 2009. Policy networks, in *European Integration Theory*, edited by A. Wiener and T. Diez. Oxford: Oxford University Press, 105–24.

Peterstorf, C., Boermans, T., Joosen, S., Jakubowska, B., Scharte, M., Stobbe, O. and Harnisch J. 2005. *Cost-Effective Climate Protection in the New EU Member States. Beyond the EU Energy Performance of Buildings Directive*. Report by ECOFYS for EurEMA. Brussels: EurEMA.

Pierson, P. 1996. The path to European integration: a historical institutional analysis. *Comparative Political Studies*, 29(2), 123–63.

Pierson, P. 2004. *Politics in Time*. Princeton, NJ: Princeton University Press.

Point Carbon 2008a. Widespread auctioning of EU allowances will cause economic hardship, 5 June.

Point Carbon 2008b. Power generators should get 500m EUAs to kick start CCS, MEPs urge, 3 July.

Point Carbon 2008c. Environment ministers divided over CCS proposals, 26 September.

Point Carbon 2008d. EU countries back using permits to fund gas capture, 14 November.

Point Carbon 2008e. MEPs blink first in Europe's CCS talks, 4 December.

Point Carbon 2011a. Brussels plans ETS 'tweaks' to avoid price slump, 11 February.

Point Carbon 2011b. EC says 22 CCS projects apply for funding, 11 March.

Pollack, M.A. 2001. International relations theory and European integration. *Journal of Common Market Studies*, 39(2), 221–44.

Pollack, M.A. 1997. Delegation, agency, and agenda setting in the European Community. *International Organization*, 51(1), 99–134.

Praetorius, B. and von Stechow, C. 2009. Electricity gap versus climate change: electricity politics and the potential role of CCS in Germany, in *Caching the Carbon – The Politics and Policy of Carbon Capture and Storage*, edited by J. Meadowcroft and O. Langhelle. Cheltenham: Edward Elgar, 125–57.

Rademaekers, K., Slingenberg, A. and Morsy, S. 2008. *Review and Analysis of EU Wholesale Energy Markets: Historical and Current Data Analysis of EU Wholesale Electricity, Gas and CO₂ Markets*. Report for the European Commission DG TREN. Rotterdam: ECORYS Nederland BV.

Ragwitz. M., Resch, G., Busch, S., Rudolf, F., Rosende, D., Held, A. and Schubert, G. 2011. *Assessment of National Renewable Energy Action Plans (interim status)*. A REPAP 2020 report. Karlsruhe/Vienna: Fraunhofer Institute Systems and Innovation Research, Karlsruhe, Germany and Vienna University of Technology, Energy Economics Group, Vienna, Austria.

Rayner, T. and Jordan, A. 2011. The United Kingdom: a paradoxical leader? in *The European Union as a Leader in International Climate Change Politics*, edited by R.K.W. Wurzel and J. Connelly. Oxford: Routledge, 95–111.

Reay, R. and C.R. Hinings. 2009. Managing the rivalry of competing institutional logics. *Organization Studies*, 30(6), 629–52.

Reiche, D. 2006. Renewable energies in the EU-accession states. *Energy Policy*, 34(3), 365–75.

Reiche, D. and Bechberger, M. 2004. Policy differences in the promotion of renewable energies in the EU member states. *Energy Policy*, 32(7), 842–49.

Reuters Planetark 2008a. Factbox – The EU's Energy and Climate Plan, 29 July.

Reuters Planetark 2008b. Setback for carbon-trapping technology in EU talks, 17 November.

Riksdagen 2002/03. *Næringsutskottets betänkande*. 2002/03:NU6. Lag om elcertificat, m.m. Stockholm: Riksdagen.

Ringius, L. 1999. Differentiation, leaders, and fairness: negotiating climate commitments in the European Community. *International Negotiation*, 4(2), 133–66.

Ringius, L. and Gupta, J. 2001. The EU's climate leadership: reconciling ambition and reality. *International Environmental Agreements*, 1(2), 281–99.

Río. P. del and Unruh, G. 2007. Overcoming the lock-out of renewable energy technologies: the case of wind and solar electricity. *Renewable and Sustainable Energy Reviews*, 11(7): 1498–1513.

Rockwool 2011. *The Rockwool Group*. [Online]. Available at: www.rockwool.com/about+the+group/the+group+in+brief [accessed: 19 June 2011].

Rowbotham, E.J. 1996. Legal obligations and uncertainties in the climate change convention, in *Politics of Climate Change*, edited by T. O'Riordan and J. Jäger. London: Routledge, 32–51.

Rowlands, I.H. 2005. The European Directive on renewable electricity: conflicts and compromise. *Energy Policy*, 33(8), 963–74.

Rueschemeyer, D. 2003. Can one or a few cases yield theoretical gains?, in *Comparative Historical Analysis in the Social Sciences*, edited by J. Mahoney and D. Rueschemeyer. Cambridge: Cambridge University Press, 305–36.

RWE 2009. *Annual Report 2008*. Essen:RWE AG.

Sæverud, I.A. and Moe, A. 2005. *Carbon Storage and Climate Change – The Case of Norway*. Lysaker, Norway: Fridtjof Nansen Institute.

Sahlin, K. and Wedlin, L. 2008. Circulating ideas: imitating, translation and editing, in *The SAGE Handbook of Organizational Institutionalism*, edited by R. Greenwood, C. Oliver, K. Sahlin and R. Suddaby. Thousand Oaks, CA: Sage, 218–42.

Schattschneider, E.E. 1960. *The Semisovereign People*. New York: Holt, Rinehart and Winston.

Schild, P.G., Klinski, M. and Grini, C. 2010. *Sammenligning og analyse av krav til energieffektivitet i bygninger i Norden og Europa* (Comparison and Analysis of Energy Performance Requirements in Buildings in the Nordic Countries and Europe). Project report 55. Trondheim: SINTEF Byggforsk.

Schimmelfenning, F. and Rittberger, B. 2006. Theories of European integration, in *European Union: Power and Policy Making*, edited by J. Richardson. Oxford: Routledge, 73–95.

Schmitter, P.C. 2010. Business and neo-corporatism, in *The Oxford Handbook of Business and Government*, edited by D. Coen, W. Grant and G. Wilson. Oxford: Oxford University Press, 248–57.

Schreurs, M. and Tiberghien, Y. 2010. European Union leadership in climate change mitigation through multilevel reinforcement, in *Global Commons, Domestic Decisions: The Comparative Politics of Climate Change*, edited by K. Harrison and L.M. Sundstrom. Cambridge, MA: MIT Press, 23–66.

Scott, W.R. 2008. *Institutions and Organizations. Ideas and Interests*. 3rd Edition. Thousands Oaks, CA: Sage.

Scrase, I. and Watson, J. 2009. CCS in the UK: squaring coal use with climate change?, in *Caching the Carbon – The Politics and Policy of Carbon Capture and Storage*, edited by J. Meadowcroft and O. Langhelle. Cheltenham: Edward Elgar, 158–85.

Sebenius, J.K. 1983. Negotiation arithmetic: adding and subtracting issues and parties. *International Organization*, 37(2), 281–316.

Sebenius, J.K. 1984. *Negotiating the Law of the Sea.* Boston/Cambridge: Harvard Economic Studies.

Sebenius, J.K. 2009. Negotiation analysis: from games to inferences to decisions to deals. *Negotiation Journal*, 25(4), 449–65.

Simon, H. [1947] 1997. *Administrative Behavior: A Study of Decision-making Processes in Administrative Organizations.* New York: Free Press.

Sims, R.E.H, Schock, R.N., Adegbululgbe, A., Fenhann, J., Konstantinaviciute, I., Moomaw, W., Nimir, H.B., Schlamadinger, B., Torres-Martínez, J., Turner, C., Uchiyama, Y., Vuori, S.J.V., Wamukonya, N. and Zhang, X. 2007. Energy supply, in *Climate Change 2007: Mitigation*, edited by B. Metz, O.R. Davidson, P.R. Bosch, R. Dave and L.A. Meyer. Cambridge: Cambridge University Press.

Skjærseth, J.B. 1994. The climate policy of the EC. *Journal of Common Market Studies*, 32(1), 21–45.

Skjærseth, J.B. 2000. *North Sea Cooperation: Linking International and Domestic Pollution Control.* Manchester: Manchester University Press.

Skjærseth, J.B. and Skodvin, T. 2009. *Climate Change and the Oil Industry: Common Problem, Varying Strategies.* Manchester: Manchester University Press.

Skjærseth, J.B. and Wettestad, J. 2002. Understanding the effectiveness of EU environmental policy: how can regime analysis contribute? *Environmental Politics*, 11(3), 99–120.

Skjærseth, J.B. and Wettestad, J. 2007. Is EU enlargement bad for environmental policy? Confronting gloomy expectations with evidence. *International Environmental Agreements*, 7(3), 263–80.

Skjærseth, J.B. and Wettestad, J. 2008. *EU Emissions Trading: Initiation, Decision-making and Implementation.* Aldershot: Ashgate.

Skjærseth, J.B. and Wettestad, J. 2010a. The EU Emissions Trading System Directive revised (Directive 2009/29/EC), in *The New Climate Policies of the European Union: Internal Legislation and Climate Diplomacy*, edited by S. Oberthür and M. Pallemaerts. Brussels: ASP Editions, 65–93.

Skjærseth, J.B. and Wettestad, J. 2010b. Fixing the EU Emissions Trading System? Understanding the post-2012 changes. *Global Environmental Politics*, 10(4), 101–23.

Skodvin, T., Gullberg, A.T., and Aakre, S. 2010. Target-group influence and political feasibility: the case of climate policy design in Europe. *Journal of European Public Policy*, 17(6), 854–73.

Slingenberg, Y. 2006. The international climate policy developments of the 1990s: the UNFCCC; the Kyoto Protocol; the Marrakech Accords and the EU Ratification Decision, in *EU Energy Law, Volume IV. EU Environmental Law, the EU Greenhouse Gas Emissions Trading Scheme*, edited by J. Delbeke, O. Hartridge, J. Lefevere, D. Meadows, A. Runge-Metzner, Y. Slingenberg, M. Vainio, P. Vis and P. Zapfel. Brussels: Claeys&Casteels, section 2.1–3.1.

Solomon, S., Qin, D., Manning, M., Chen, Z., Marquis, M., Averyt, K.B., Tignor, M. and Miller, H.L. (eds) 2007. Climate Change 2007 – The Physical Science Basis. Working Group I contribution to the Fourth Assessment Report of the Intergovernmental Panel on Climate Change. New York, NY: Cambridge University Press.

Sorell, S. (ed.) 2003. Interaction in EU Climate Policy. SPRU. Final report. Project No: EVK2-CT-2000-0067.

Spiegel 2011. 408 mögliche CO2-Endlager in Deutschland. Der Spiegel [Online]. Available at: www.spiegel.de/wissenschaft/technik/0,1518,745191,00.html [accessed: 14 February 2011].

Stavins, R.N. 2002. Lessons from the American experiment with market-based environmental policies, in *Market-Based Governance: Supply Side, Demand Side, Upside, and Downside*, edited by J.D. Donahue and J.S. Nye, Jr. Washington, DC: Brookings Institution Press, 173–201.

Stephens, J.C. 2009. *Technology leader, policy laggard: CCS development for climate mitigation in the US political context, in Caching the Carbon – The Politics and Policy of Carbon Capture and Storage*, edited by J. Meadowcroft and O. Langhelle. Cheltenham: Edward Elgar, 22–49.

Stern, N. 2006. *Stern Review on the Economics of Climate Change*. London: Her Majesty's Treasury.

Stokke, O.S. 2001. The Interplay of International Regimes: Putting Effectiveness Theory to Work. FNI Report 14/2001. Lysaker, Norway: Fridtjof Nansen Institute.

Stokke, O.S., Hovi, J. and Ulfstein, G. 2005. *Implementing the Climate Regime: International Compliance*. London: Earthscan

Strang, D. and Meyer, J.W. 1993. Institutional conditions for diffusion. *Theory and Society*, 22, 487–511.

Streeck, W. and Thelen, K. 2005. *Beyond Continuity*. Oxford: Oxford University Press.

Tarlock, A.D. 1992. The role of non-governmental organizations in the development of international environmental law. *Chicago-Kent Law Review*, 68(61), 61–76. [Online]. Available at: http://works.bepress.com/dan_tarlock/58 [accessed: 30 March 2012].

ten Brink, P. 2010. Mitigating CO2 emissions from cars in the EU (Regulation (EC) No. 443/2009), in *The New Climate Policies of the European Union*, edited by S. Oberthur and M. Pallemaerts. Brussels: VUB Press, 179–211.

Thomsen, K.E. and Aggerholm, S. 2009. Denmark: Impact, Compliance, and Control of Legislation. ASIEPI information paper P175. [Online]. Available at: www.buildup.eu/publications/7047 [accessed: 14 February 2012].

Thornton, P. 2004. *Markets from Culture*. Stanford, CA: Stanford Business Books.

Thornton, P.H. and Ocasio, W. 2008. Institutional logics, in *The SAGE Handbook of Organizational Institutionalism*, edited by R. Greenwood, C. Oliver, K. Sahlin and R. Suddaby. Thousand Oaks, CA: Sage, 99–129.

Tolbert, P.S. and Zucker, L. G. 1983. Institutional sources of change in the formal structure of organizations: the diffusion of civil service reform. *Administrative Science Quarterly*, 28 (1), 22–39

Total 2009. Carbon Dioxide Capture and Geological Storage. [Online]. Available at: www.total.com/en/challenges/carbon-dioxide-capture-and-geological-storage/total-s-commitment/research-programs-and-industrial-projects-940766.html [accessed: 5 February 2010].

Trondal, J. 2010. *An Emergent European Executive Order*. Oxford: Oxford University Press.

UNFCCC. 2009. Draft decision -/CP.15 Proposal by the President Copenhagen Accord. FCCC/CP/2009/L.7 18 December 2009. United Nations Framework Concention on Cliamte Change. Conference of the Parties, Copenhagen.

Urge-Vorsatz, D., Miladinova, G. and Paizs, L. 2006. Energy in transition: from the iron curtain to the European Union. *Energy Policy*, 34(15), 2279–97.

Van Asselt, H. 2010. Emissions trading: the enthusiastic adoption of an 'alien' instrument?, in *Climate Change Policy in the European Union. Confronting the Dilemmas of Mitigation and Adaptation*, edited by A. Jordan, D. Huitema, H.V. Asselt, T. Rayner and F. Berkhout. Cambridge: Cambridge University Press, 125–45.

Vattenfall, 2009. *Annual Report 2008*. Stockholm:Vattenfall

Vergragt, P.J. 2009. CCS in the Netherlands: glass half empty or half full?, in *Caching the Carbon – The Politics and Policy of Carbon Capture and Storage*, edited by J. Meadowcroft and O. Langhelle. Cheltenham: Edward Elgar, 186–210.

Victor, D.G. and House, J.C. 2006. BP's emissions trading system. *Energy Policy*, 34(15), 2100–2112.

Visier, J.C., Thomsen K.E. and Johanssen, G. 2003. Energy Performance of Buildings. ENPER-TEBUC report. St-Stevens-Woluwe: Belgian Building Research Institute.

Voss, J.P. 2007. Innovation processes in governance: the development of 'emissions trading' as a new policy instrument. *Science and Public Policy*, 34(5), 329–43.

Wagner, J.P. 1997. The climate change policy of the European Union, in *International Politics of Climate Change*, edited by G. Fermann. Oslo: Scandinavian University Press, 297–340.

Wallace, H. and Wallace, W. (eds) 2000. Policy-Making in the European Union. Oxford: Oxford University Press.

Wallace, H. and Wallace W. 2006. Overview: the European Union, politics and policy-making, in *Handbook of European Union Politics*, edited by K.E. Jørgensen, M.A. Pollack and B. Rosamond. Thousand Oaks, CA: Sage, 339–58.

Watanebe, R. 2011. *Climate Policy Changes in Germany and Japan*. London: Taylor & Francis.

Wettestad, J. 1999. Increasing concern and improving design: the Oslo and Paris Conventions on Marine Pollution in the North-East Atlantic, in *Designing Effective Environmental Regimes: The Key Conditions*. Cheltenham: Edward Elgar, 43–84.

Wettestad, J. 2000. The complicated development of EU climate policy, in *Climate Change and European Leadership*, edited by M. Grubb and J. Gupta. Dordrecht: Kluwer Academic Publishers, 25–47.

Wettestad, J. 2001. The ambiguous prospects for EU climate policy – a summary of options. *Energy and Environment*, 12(2&3), 139–67.

Wettestad, J. 2002. *Clearing the Air – European Advances in Tackling Acid Rain and Atmospheric Pollution*. Aldershot: Ashgate.

Wettestad, J. 2005. The making of the 2003 EU Emissions Trading Directive: ultra-quick process due to entrepreneurial proficiency? *Global Environmental Politics*, 5(1), 1–24.

Wettestad, J. 2009a. EU energy-intensive industries and emission trading: losers becoming winners? *Environmental Policy and Governance*, 19(5), 309–20.

Wettestad, J. 2009b. European climate policy: towards centralised governance? *Review of Policy Research*, 26(3), 311–29.

Wettestad, J. 2009c. Interaction between EU carbon trading and the international climate regime: synergies and learning. *International Environmental Agreements*, 9, 393–408.

Wettestad, J. 2011. EU emissions trading: achievements and challenges, in *Towards A Common European Union Energy Policy: Problems, Progress and Prospects*, edited by J.S. Duffield and V.L. Birchfield. New York: Palgrave, 87–113.

Wettestad, J. and Hals Butenschøn, S. 2000. The Increasing British Climate Ambitiousness: A Mere Reflection of 'The Dash for Gas'? FNI Report 3/2000. Lysaker, Norway: Fridtjof Nansen Institute.

Wettestad, J. and Sæverud, I. A. 2005. Implementing EU Emissions Trading: Institutional Misfit? FNI Report 10/2005. Lysaker, Norway: Fridtjof Nansen Institute.

Wettestad, J., Eikeland, P.O. and Nilsson, M. 2012. EU climate and energy policy: a hesitant supranational turn? *Global Environmental Politics*, 12(2), forthcoming May 2012.

Woods, P. 2010. Implementation of the EPBD in England and Wales, Scotland and Northern Ireland. Status in November 2010. [Online]. Available at: www. epbd-ca.org/Medias/Pdf/country_reports_14-04-2011/England_and_Wales_Scotland_and_Northern_Ireland.pdf [accessed: 14 February 2012].

Wurzel, R.K.W. 2002. *Environmental Policy-making in Britain, Germany and the European Union*. Manchester: Manchester University Press.

Wurzel, R.K.W. 2008. The Politics of Emissions Trading in Britain and Germany. Report for the Anglo-German Foundation for the Study of Industrial Society. London: Anglo-German Foundation.

Wurzel, R.K.W. and Connelly, J. (eds) 2011a. *The European Union as a Leader in International Climate Change Politics*. Oxford: Routledge.

Wurzel, R.K.W. and Connelly, J. 2011b. Introduction: European Union political leadership in international climate change politics, in *The European Union as*

a Leader in International Climate Change Politics, edited by R.K.W. Wurzel
and J. Connelly. Oxford: Routledge.

Wurzel, R.K.W. and Connelly, J. 2011c. Environmental NGOs: taking a lead?, in
The European Union as a Leader in International Climate Change Politics,
edited by R.K.W. Wurzel and J. Connelly. Oxford: Routledge, 214–29.

Young, O.R. 2002. *The Institutional Dimensions of Environmental Change: Fit,
Interplay, and Scale*. Cambridge MA: MIT Press.

Young, A.R. 2006. *The politics of regulation and the internal market, in Handbook
of European Union Politics*, edited by K.E. Jørgensen, M.A. Pollack and B.
Rosamond. Thousand Oaks, CA: Sage, 373–94.

ZEP 2006. The European Technology Platform for Zero Emission Fuel Power
Plants (ZEP), Strategic Deployment Document. Brussels: ZEP.

List of Interviews

Ahluwalia, Kavita, European Parliament, Brussels, 28 May 2009 (interview conducted together with Jon B.Skjærseth).

Andersson, Bosse, Vattenfall, Stockholm, 13 March 2009.

Baumann, Knut, Federation of Norwegian Industries, Oslo, 24 January 2008.

Bergflødt, Lise, Skanska, Oslo, 7 November 2008.

Blanken, Joris van den, Greenpeace, Brussels, 22 June 2009.

Bornkamm, Oliver, German Brussels Delegation, Brussels, 24 June 2009.

te Bos, Jan, EURIMA, Brussels, 25 June 2009.

Bowie, Randall, Rockwool, Brussels, 26 February 2008; 26 January 2011.

Brockett, Scott, DG Environment (later DG Climate), European Commission, Brussels, 26 January 2011.

Carl, Mogens Peter, former director of DG Environment and advisor to the French Presidency, autumn 2008, Skype interview, 28 February 2012.

Cazes, Bertrant, Glass for Europe, Brussels, 28 January 2011.

Dahl, Agnethe, Norwegian Ministry of the Environment, Oslo, 15 June 2009.

Davies, Chris, European Parliament, Brussels, 25 May 2011.

Delbeke, Jos, DG Climate, European Commission, Brussels, 27 January 2011.

Dokka, Tor Helge, Sintef Building and Infrastructure, Oslo, 5 October 2008.

Doppelhammer, Martina, DG Climate, European Commission, Brussels, 28 January 2011.

Edwards, Sorcha, CECODHAS (Housing Europe), Brussels, 27 January 2011.

Egenhofer, Christian, Center for European Policy Studies (CEPS), 31 October 2008, Brussels, 27 January 2011.

Ehrenberg, Joachim, DG Enterprise, European Commission, Brussels, 30 October 2008.

Engebretsen, Marit, Norwegian Ministry of Energy, Brussels Delegation, Brussels, 23 June 2009.

Eriksen, Henrik, Norwegian Ministry of the Environment, Brussels Delegation, Brussels, 23 June 2009.

Fagernes, Ronald, Federation of Norwegian Industries, Oslo, 24 January 2008.

Faraday, Frank, European Construction Industry Federation (FIEC), Brussels, 28 February 2008.

Foquet, Dörte, European Renewable Energies Federation (EREF), Brussels, 28 February 2008 (interview conducted together with Karoline Flåm).

Frisvold, Paal, Bellona, Brussels, 23 July 2009.

Geiss, Jan, European Forum for Renewable Energy Sources (EUFORES), Brussels, 24 January 2011.

Goodall, John, European Construction Industry Federation (FIEC), Brussels, 28 February 2008.

Hall, Fionna, European Parliament, Brussels, 25 January 2011.

Hedenstrøm, Claes,Vattenfall, Stockholm, 13 March 2009.

Hercsuth, Andrea, European Commission, DG Transport and Energy, Brussels, 27 February 2008.

Holm, Marius, Bellona, Oslo, 10 June 2009.

Jeekel, Robert, Eurometaux, Brussels, 27 May 2009 (interview conducted with Jon B. Skjærseth).

Kopczynski, Olaf, Polish Brussels Delegation, Brussels, 26 January 2011.

Koskimaki, Pirjo-Liisa, European Commission, DG Transport and Energy/DG Energy, Brussels, 25 June 2009; 25 January 2011.

Kroepelien, Knut, Norwegian Ministry of the Environment, Brussels Delegation, Brussels, 23 June 2009.

Kumar, Sanjeev, WWF, Brussels, 31 October 2008; 22 June 2009.

Lipponen, Juha, EURELECTRIC, Brussels, 25 February 2008; 24 June 2009 (first interview conducted with Karoline Flåm, Per Ove Eikeland and Måns Nilsson).

Mensink, Marco, European Pulp and Paper Federation (CEPI), Brussels, 26 February 2008; 28 January 2011 (second interview conducted with Lars Gulbrandsen and Liv Arntzen Løchen).

Møller, Ulf, Statnett (Norwegian Transmission System Operator), Oslo, 17 January 2008.

Piel, Elo, Euroheat and Power, Brussels, 29 March 2008 (conducted with Karoline Flåm).

Pitkethly, Erik, UK Brussels Delegation, Brussels, 26 January 2011.

Rushe, Tim Maxian, European Commission, DG Transport and Energy, Brussels, 27 February 2008 (interview conducted with Karoline Flåm, Per Ove Eikeland and Måns Nilsson).

Schaefer, Oliver, European Renewable Energy Council (EREC), Brussels, 25 June 2009.

Schneider, Norbert, EON Brussels office, Brussels, 25 May 2011.

Scott, Jesse, E3G, Brussels, 25 May 2011.

Scowcroft, John EURELECTRIC, Brussels, 25 February 2008.

Seinen, Anne-Theo, European Commission, DG Environment, Brussels, 26 February 2008.

Sharpe, Dale, UK Brussels Delegation, Brussels, 23 June 2009.

Slingenberg, Yvon, European Commission, DG Climate/DG Environment, Brussels, 31 October 2008; 27 January 2011(second interview conducted with Lars Gulbrandsen and Liv Arntzen Løchen).

Solheim, Marit, Norwegian Ministry of the Environment, Oslo, 15 June 2009.

Steen, Hans van, European Commission, DG Transport and Energy, Brussels, 23 June 2009.

Stiansen, Peer, Norwegian Ministry of the Environment, 30 April 2008; 22 April 2009.

Sundsbø, Svein, Federation of Norwegian Industries, Oslo, 24 January 2008.

Ticau, Adriana Silvia, European Parliament, Brussels, 24 June 2009.

Ulseth, Oluf, Statkraft, Lysaker (Norway), 8 October 2008.

Veum, Karina, European Commission, DG Transport and Energy, telephone interview, 28 March 2008.

Vis, Peter, DG Climate/DG TREN, European Commission, Brussels, 22 June 2009; 26 January 2011.

Warren, Andrew, Eurace, Brussels, 25 June 2009.

Westgård, Geir, Statoil Brussels office, Brussels, 22 June 2009.

Wyns, Tomas, CAN Europe, Brussels, 30 October 2008; 26 January 2011(second interview conducted with Lars Gulbrandsen and Liv Arntzen Løchen).

Index